USE OF SOMATOTROPIN IN LIVESTOCK PRODUCTION

A seminar on Use of Somatotropin in Livestock Production, held in Brussels, 27–29 September 1988, as part of the EC Programme for the Coordination of Agricultural Research.

Sponsored by the Commission of the European Communities, Directorate-General for Agriculture, Coordination of Agricultural Research Division.

USE OF SOMATOTROPIN IN LIVESTOCK PRODUCTION

Edited by

K. SEJRSEN

M. VESTERGAARD

and

A. NEIMANN-SØRENSEN

*National Institute of Animal Science,
Foulum, 8830 Tjele, Denmark*

ELSEVIER APPLIED SCIENCE
LONDON and NEW YORK

ELSEVIER SCIENCE PUBLISHERS LTD
Crown House, Linton Road, Barking, Essex IG11 8JU, England

Sole Distributor in the USA and Canada
ELSEVIER SCIENCE PUBLISHING CO., INC.
655 Avenue of the Americas, New York, NY 10010, USA

WITH 100 TABLES AND 33 ILLUSTRATIONS

© 1989 ECSC, EEC, EAEC, BRUSSELS AND LUXEMBOURG

British Library Cataloguing in Publication Data
Use of somatotropin in livestock production.
1. Livestock. Production. Growth hormones
I. Sejrsen, K. II. Vestergaard, M. III. Neimann-Sørensen, A.
636.08′52

ISBN 1-85166-386-X

Library of Congress CIP data applied for

Publication arrangements by Commission of the European Communities, Directorate-General Telecommunications, Information Industries and Innovation, Scientific and Technical Communication Service, Luxembourg

EUR 11881

LEGAL NOTICE
Neither the Commission of the European Communities nor any person acting on behalf of the Commission is responsible for the use which might be made of the following information.

No responsibility is assumed by the Publisher for any injury and/or damage to persons or property as a matter of products liability, negligence or otherwise, or from any use or operation of any methods, products, instructions or ideas contained in the material herein.

Special regulations for readers in the USA

This publication has been registered with the Copyright Clearance Center Inc. (CCC), Salem, Massachusetts. Information can be obtained from the CCC about conditions under which photocopies of parts of this publication may be made in the USA. All other copyright questions, including photocopying outside the USA, should be referred to the publisher.

All rights reserved. No part of this publication may be reproduced, stored in a retrieval system, or transmitted in any form or by any means, electronic, mechanical, photocopying, recording, or otherwise, without the prior written permission of the publisher.

Printed in Great Britain by Galliard (Printers) Ltd, Great Yarmouth

PREFACE

It has been known for decades that growth hormone—somatotropin—stimulates growth and lactation in farm animals, but it has not been possible to take advantage of this knowledge in animal production, because naturally produced somatotropin is only available in small amounts. Thanks to the development of recombinant DNA technology, somatotropin can now be produced in quantities sufficient for practical use, and commercial companies have applied for permission to market somatotropin produced by this technology in the European Community. Approval of products for enhancement of production efficiency is usually based on their efficiency, quality and safety. Somatotropin represents a special case because it is one of the first products of biotechnology approaching practical application in animal agriculture.

This publication contains papers presented at a seminar held in Brussels on 27–29 September 1988 under the auspices of the Commission of the European Communities. The objective of the seminar was to review and discuss scientific information relevant to the decision on whether to approve or reject the applications. Reviews of the available scientific information on mechanisms of somatotropin action and influences of somatotropin on lactation, growth, product quality and animal health and welfare were presented. A session covered the socio-economic effects at farm and national as well as Community level and a paper discussed the possible future alternatives to somatotropin. Based on the discussions and the information presented, a summary report with concluding remarks was prepared at and shortly after the seminar.

The seminar was organised by Dr K. Sejrsen and Professor A. Neimann-Sørensen with assistance from Dr J. Connell and Dr M. Vestergaard. The seminar was opened by Dr B. Hogben and the sessions were chaired by Dr K. Sejrsen, Professor A. Neimann-Sørensen, Professor J. Fabry, Professor B. Hoffmann and Dr M. Tracey. The discussions on lactation, growth and product quality was opened by Professor D. E. Baumann, Professor F. Ellendorff and Professor B. Hoffmann. The general discussion was chaired by Professor A. Neimann-Sørensen. Dr W. F. Raymond acted as rapporteur and prepared the summary report with Professor A. Neimann-Sørensen.

CONTENTS

Preface v

BIOLOGY OF SOMATOTROPIN

Mechanism of action of bovine somatotropin in increasing milk secretion in dairy ruminants 1
   C. G. Prosser and T. B. Mepham

Variation of BST and IGF-I concentrations in blood plasma of cattle . 18
   D. Schams, U. Winkler, M. Theyerl-Abele and A. Prokopp

Influence of somatotropin on metabolism 31
   R. G. Vernon

Alternatives to growth hormone for the manipulation of animal performance 51
   D. J. Flint

USE OF SOMATOTROPIN FOR MILK PRODUCTION

Long-term effects of recombinant bovine somatotropin (rBST) on dairy cow performances: a review 61
   Y. Chilliard

A review of the influence of somatotropin on health, reproduction and welfare in lactating dairy cows 88
   R. H. Phipps

Influences of somatotropin on evaluation of genetic merit for milk production 120
   H. O. Gravert

USE OF SOMATOTROPIN FOR GROWTH

Effects of administration of somatotropin on growth, feed efficiency and carcass composition of ruminants: a review 132
   W. J. Enright

Use of somatotropin in livestock production: growth in pigs . . . 157
   T. J. Hanrahan

INFLUENCE OF SOMATOTROPIN ON PRODUCT QUALITY

Milk from BST-treated cows: its quality and suitability for processing . 178
   G. van den Berg

Somatotropin and related peptides in milk 192
   D. Schams

Effects of administration of somatotropin on meat quality in ruminants: a review 201
   P. Allen and W. J. Enright

Effects of pST on carcass composition and meat quality 210
   M. Henning, E. Hüster, R. E. Ivy, E. Kallweit and F. Ellendorff

SOCIO-ECONOMIC EFFECTS OF SOMATOTROPIN APPLICATION

Potential farm level and dairy sector impact of the use of bovine somatotropin (BST) in the Federal Republic of Germany . . . 212
   J. Zeddies and R. Doluschitz

Application of bovine somatotropin (BST) in milk production: possible benefits and problems 251
   H. Glaeser and J. Gay

Application of somatotropin in meat production: impact at Commission level 256
   R. Nagel

SHORT COMMUNICATIONS

The effects of treatment of dairy cows of different breeds in a second lactation with recombinantly derived bovine somatotropin in a sustained delivery vehicle 262
   J. K. Oldenbroek, G. J. Garssen, L. J. Jonker and J. I. D. Wilkinson

Recombinant somatotropin—a survey on a 2 years experiment with dairy cows 267
    P. Lebzien, K. Rohr, R. Daenicke and D. Schlünsen

Effect of slow-released somatotropin on dairy cow performances . . 269
    R. Vérité, H. Rulquin and Ph. Faverdin

BST effects on metabolism parameters in dairy cows—experimental data 274
    E. Farries

Effect of bovine somatotropin on milk yield and milk composition in periparturient cows experimentally infected with *Escherichia coli* . . 277
    C. Burvenich, G. Vandeputte-van Messom, E. Roets, J. Fabry and A.-M. Massart-Leen

Bovine somatotropin—the practical way forward related to animal welfare 281
    W. Vandaele

First simulation results on the impact of the use of BST on genetic gains for milk yield 285
    J. J. Colleau

Variability of responsiveness to growth hormone in ruminants: nutrient interactions 286
    J. M. Pell, M. Gill and D. E. Beever

Effect of recombinant porcine somatotropin (rpST) on fattening performance and meat quality of three genotypes of pigs slaughtered at 100 and 140 kg 288
    P. van der Wal, E. Kanis, W. van der Hel, J. Huisman and M. W. A. Verstegen

Serum somatotropin values in boars and barrows, implanted with anabolic steroids and compared to untreated controls 296
    R. O. De Wilde, H. Deschuytere and M. Corijn

Consequences of dietary lipid feeding in periparturient swine on endogenous growth hormone secretion and subsequent litter performance . . . 300
    G. Janssens and R. De Wilde

Fish growth hormones 304
    A. Renard, C. Lecomte, F. Rentier and J. A. Martial

The effect of BST on production of milk intended for the production of Grana cheese: technical and economic aspects 307
    G. Piva, F. Masoero and A. Lazzari

SUMMARY AND CONCLUDING REMARKS

Use of somatotropin in livestock production in the European Community:
Seminar summary and concluding remarks 312
    W. F. Raymond and A.Neimann-Sørensen

List of Participants 327

MECHANISM OF ACTION OF BOVINE SOMATOTROPIN IN INCREASING MILK SECRETION IN DAIRY RUMINANTS

C.G. Prosser[1] and T.B. Mepham[2]

1. AFRC Institute of Animal Physiology and Genetics Research, Babraham, Cambridge CB2 4AT, U.K.
2. Department of Physiology and Environmental Science, University of Nottingham, School of Agriculture, Sutton Bonington, Loughborough, U.K.

ABSTRACT

This paper reviews evidence relating to the physiological mechanism by which administration of exogenous bovine somatotropin (derived from pituitary glands or by recombinant DNA technology) elicits its galactopoietic effects in dairy ruminants. Three components of the response have been identified: changes in whole-body nutrient utilization; cardiovascular changes which result in increased nutrient supply to mammary tissue; and changes in the secretory capacity of mammary tissue. The roles of somatotropin, its putative mediator (insulin-like growth factor I) and other endocrine factors in effecting these changes are reviewed.

INTRODUCTION

Bovine somatotropin (growth hormone) is a single chain polypeptide, of 191 amino acids, produced by the anterior pituitary gland. The amino acid composition of somatotropins from various species shows appreciable differences (Wallis, 1989). For example, human and bovine somatotropin differ by 35% but ovine and bovine somatotropins are almost identical. Within a species, somatotropin extracted from the pituitaries also exhibits some degree of micro-heterogeneity, the physiological significance of which is not fully understood (Wallis, 1989).

Somatotropin was first known for its ability to enhance body weight gain and increase protein accretion (Evans and Simpson, 1931; Lee and Schaffer, 1934), but in ruminants it also possesses galactopoietic properties (Asimov and Krouze, 1937; Young, 1947). Since the advent, in the 1970's, of recombinant DNA technology, sufficient quantities of individual molecular species of somatotropin have been produced in a highly purified form to permit assessment of their efficacy and mode of action in growth and lactation. In recent years a number of reviews have appeared which deal with both of these subjects. Rather than simply attempting to update this documentation, our aim here is to identify what might prove to be key data in describing the mechanism of action of somatotropin in enhancing milk secretion in dairy ruminants. The effects of somatotropin on whole animal growth are discussed in comprehensive reviews of this area by Boyd and Bauman (1988) and also Etherton (1989).

ENDOCRINE CHANGES

The major consistent changes in concentrations of hormones/growth factors during somatotropin treatment are those of somatotropin itself and insulin-like growth factor I (IGF-I). The 3 to 4-fold increases in plasma concentrations of IGF-I observed in late lactating cows (Davis et al., 1987; Prosser, Fleet and Corps, 1988) and sheep (Fleet et al., 1988) are consistent with somatotropin control of circulating levels of IGF-I in non-lactating animals (Nissley and Rechler, 1985). IGF-I production is also acutely sensitive to the nutritional status of the animal. Reduced levels of plasma IGF-I are observed even in the presence of elevated somatotropin in fasted humans (Merimee, Zapf and Froesch, 1982) and rats (Maes et al., 1984; Maiter et al., 1988) and in cows in energy deficit at peak lactation (Ronge et al., 1988). The ability of exogenous somatotropin to stimulate circulating IGF-I in lactating ruminants in different states of energy balance has not been documented.

No significant alterations in the concentrations of ACTH, prolactin, TSH, LH or FSH have been reported (Peel et al., 1983; Schemm and Deaver, 1988; Lanza et al., 1988; Peel, Eppard and Hard, 1989). These data suggest that prolonged treatment with somatotropin does not alter anterior pituitary function with respect to these hormones, although studies of the function of the hypothalamic-pituitary axis in relation to endogenous release of somatotropin during chronic treatment of lactating animals are lacking. Concentrations of glucagon and tri-iodothyroxine are also unchanged but there are reports of a small increase in concentrations in blood of thyroxine (Peel et al., 1983; Bitman et al., 1984; Lanza et al., 1988; Peel, Eppard and Hard, 1989).

Increases in circulating concentrations of insulin have sometimes, but not always, been observed during somatotropin treatment (Bines, Hart and Morant, 1980; Peel et al., 1983; Eppard, Bauman and McCutcheon, 1985; McDowell et al., 1987a; Fullerton et al., 1989; Soderholm et al., 1988). This variability may relate to the manner in which animals in differing energy states respond to treatment or to sampling regimens which do not adequately account for acute fluctuations in circulating concentrations.

Somatotropin is antagonistic to insulin action in peripheral tissues (Davidson, 1987), so an increase in insulin release may serve to overcome this resistance. In the rat, somatotropin can stimulate isolated pancreatic islets to synthesize and release insulin (Schatz et al., 1973), but whether this is true of ruminants is not known. Somatotropin also influences insulin secretion through changes in nutrient concentrations in blood. In non-ruminants glucose is a key regulator of insulin secretion whereas in ruminants volatile fatty acids, because of their greater concentrations, may be more important. Whether the response of the pancreas of ruminants to insulinotropic agents is altered during somatotropin treatment is not known.

RECEPTORS FOR SOMATOTROPIN

Biological responses to agonists such as hormones require the presence of specific receptors on the surface of target cells. In non-ruminants, somatogenic receptors have been detected in liver (Tsuchima and Friesen, 1973; Hughes, 1979), adipocytes (Di Girolamo et al., 1986), fibroblasts (Murphy, Vrhovsek and Lazarus, 1983), lymphocytes (Lesniak and Roth, 1976) and chondrocytes (Isaksson et al., 1987). There is little information on the distribution and nature of somatogenic receptors in different tissues of ruminants. The evidence that is available suggests there are at least two classes of receptors in ovine (Gluckman, Butler and Elliot, 1983) and bovine (Breier, Gluckman and Bass, 1988) liver, based on the affinity of the receptors for bovine somatotropin. The function of the two classes of receptor is not clear, but it seems that the production of IGF-I, for instance, may be correlated with the presence of the high affinity receptor (Maes et al., 1984; Gluckman and Breier, 1989).

The number and function of somatogenic receptors in liver is regulated by a variety of factors including somatotropin, steroids and nutrition (Baxter, Zaltsman and Turtle, 1984; Maiter et al., 1988; Gluckman and Breier, 1989). Since the overall response of the animal to exogenous somatotropin treatment depends on the relative abundance of functional receptors in target tissues, the importance of the nutritional status of the animal in determining the response can readily be seen. This may provide the underlying physiological mechanism by which the galactopoietic response is constrained by the level of management of treated animals.

While mammary tissue from sheep and cows binds human somatotropin and bovine prolactin (Akers and Keys, 1984), there appears to be little specific binding of bovine somatotropin (Akers, 1985), indicating the presence of lactogenic but not somatogenic receptors in mammary gland of lactating ruminants. Beckers (1987) observation of somatogenic receptors in mammary tissue taken from pregnant cows may represent binding to non-epithelial components. The more usually reported inability to detect significant specific binding of somatotropin to lactating mammary tissue suggest somatotropin may not act directly on the mammary epithelium to alter milk yield.

NUTRIENT PARTITIONING

When lactating cows are treated with somatotropin milk yield is increased immediately whereas the corresponding increase in voluntary feed intake does not occur until at least the 5th to 7th week of treatment (Bauman et al., 1985; Elvinger et al., 1988; Soderholm et al., 1988). Thus in the initial stages of treatment the requirements for the extra nutrients must be met by mobilization of body stores.

In 1980 Bauman and Currie put forward a scheme whereby a homeorhetic signal, or signals, would induce 'orchestrated changes' in nutrient utilization to support the increased demands of the mammary gland during normal lactation. According to this scheme, during early lactation, nutrients are mobilized from body stores laid down during the dry period and re-partitioned towards the mammary gland. Perhaps the most convincing evidence for somatotropin acting as a homeorhetic signal to alter nutrient partitioning was obtained by McDowell et al. (1987b). In their study, the uptake of key metabolites was measured simultaneously in hind limb muscle and mammary gland of somatotropin treated cows. Mammary uptake of glucose and non-esterified fatty acids (NEFA's) was enhanced while that of muscle was reduced by somatotropin.

The lag in voluntary feed intake in the initial stages of treatment also means that cows are generally in negative energy balance, especially if the milk yield response is high (Eppard, Bauman and McCutcheon, 1985; Bauman et al., 1985; Elvinger et al., 1988; Soderholm et al., 1988). Eventually, as feed intake increases the animals attain positive energy status. Therefore the adaptations in whole body metabolism which support the extra milk yield and the factors that control these events must vary during treatment. This would seem to account for the often conflicting reports of the changes in metabolite levels during treatment.

A further point that is immediately obvious when examining the galactopoietic effect of administered somatotropin is that, apart from a minor fall in milk protein and rise in milk fat content which sometimes occur, the concentrations of the major nutrients of milk are usually unaltered (Peel et al., 1983; Bitman et al., 1984; Bauman et al., 1985; Eppard, Bauman and McCutcheon, 1985; McDowell et al., 1987a; Soderholm et al., 1988; Elvinger et al., 1988; Fullerton et al., 1989; Peel, Eppard and

Hard, 1989). The increase in lactose output during somatotropin treatment is met by increased diversion of glucose to the mammary gland. It has been suggested that additional glucose required arises from its decreased oxidation in peripheral tissues and increased hepatic gluconeogenesis from propionate (Pocius and Herbein, 1986) and/or glycerol (Peel and Bauman, 1987). Circulating concentrations of glucose are sometimes (McDowell et al., 1987a; Fleet et al., 1988; Soderholm et al., 1988), but not always (Peel et al., 1983; Eppard, Bauman and McCutcheon, 1985; Fullerton et al., 1989), increased by somatotropin, reflecting the balance between its production and utilization.

Concentrations of protein in milk usually remain constant during somatotropin treatment, but sometimes decrease slightly (Peel et al., 1983; McDowell et al., 1987a). Presumably this results from an insufficient intracellular supply of amino acids for milk protein synthesis rather than decreased mRNA for milk proteins or the capacity to synthesize and secrete the proteins, as the decrease seems to occur when animals have a reduced energy balance. Very few comprehensive analyses of milk protein composition in treated animals have been reported. One report showed increased α-lactalbumin in milk, which exceeded the response in other milk proteins (Eppard et al., 1985), whereas another showed no change (Lynch et al., 1988). Increased concentrations of IGF-I in milk have also been observed (to be discussed later). Changes in accretion rate of muscle protein and oxidation and utilization of amino acids as gluconeogenic substrates are probably all involved in 'sparing' amino acids for synthesis of milk proteins. However, studies of the kinetics of amino acid metabolism during somatotropin treatment are not yet complete.

One of the most potent effects of somatotropin on metabolism relates to its effects on adipose tissue. The response is twofold, i.e. decreased lipid synthesis, which thus 'spares' acetate and glucose, and enhanced lipolysis, releasing NEFA's (Vernon and Flint, 1989). The net result is that cows receiving somatotropin have lower body fat than untreated animals, a fact which may account for the reported decrease in live weight (Soderholm et al., 1988). In most studies there was an increase in plasma concentrations of NEFA's when treated cows were in negative energy balance, but no change when they were in positive energy balance (Peel et al., 1983; Bitman et al., 1984; Eppard, Bauman and McCutcheon, 1985; Soderholm et al., 1988). In contrast, McDowell et al. (1987a) reported an increase in plasma NEFA's when treated cows were in positive energy balance. The discrepancy is probably not attributable to differences in treatments since similar treatment periods and doses were given. Generally, when plasma NEFA's increase during somatotropin treatment milk fat content also increases. The composition of milk fat in these cases shifts to a greater proportion of long chain fatty acids, indicating that more of the lipids in milk are derived from the plasma (Bitman et al., 1984).

The data from the study by McDowell et al. (1987a) warrant close scrutiny because they offer important insights into the complexities of the metabolic effects of somatotropin. In this study Friesian cows at peak (~ 40 days) and mid-lactation (~ 130 days) were treated for 6 days with pituitary-derived bovine somatotropin. Milk yield increased 6 and 14% at peak and mid-lactation respectively. Cows were in negative energy balance at peak lactation and positive energy balance at mid-lactation both before and during treatment. Cows at peak lactation exhibited increased plasma glucose concentrations and irreversible losses and no changes in concentrations of acetate or urea, but small increases in their irreversible losses. There was no change in plasma NEFA concentrations but their irreversible loss decreased. Cows at mid-lactation had increased plasma concentrations of glucose and NEFA's and decreased acetate and urea. Irreversible losses of glucose and acetate were not altered while that of urea decreased slightly and that of NEFA's increased. The differences reflect increased utilization of NEFA's as a metabolizable fuel at mid-lactation, sparing glucose and acetate. At peak lactation glucose and acetate appeared to be the more important sources of energy. Increases in plasma concentrations of 3-hydroxybutyrate, which occurred at mid-lactation, support this hypothesis.

Studies of this nature emphasize the need to consider the metabolic changes in total, rather than alterations in the circulating levels of individual components. They also demonstrate the metabolically diverse manner in which animals respond to somatotropin treatment in supporting increased milk yield.

CARDIOVASCULAR EFFECTS

A number of studies have now demonstrated an increased rate of blood flow to the mammary gland during somatotropin treatment (Mepham et al., 1984; Davis et al., 1988a; Fullerton et al., 1989; Prosser, Fleet and Corps, 1988). This provides a further means by which nutrient supply to the gland may be regulated. Indeed in some instances it would appear that the increase in mammary uptake of some precursors is accounted for solely by an increase in mammary blood flow (Davis et al., 1988b; Fullerton et al., 1989).

Short-term (4 days) treatment of sheep (Fleet et al., 1988) and cows (Davis et al., 1988a) with somatotropin also increased cardiac output. As with the increase in cardiac output which occurs during early lactation in the rat (Hanwell and Linzell, 1973), a greater proportion of the increased cardiac output which results from somatotropin treatment is diverted toward the mammary glands (Davis et al., 1988a).

The precise role that the increase in mammary blood flow plays in the galactopoietic response to somatotropin remains to be elucidated. For example, it is not known whether the increase in mammary blood flow is the cause or consequence of the increase in mammary activity. In a study in somatotropin-treated goats, on some occasions increases in mammary blood flow occurred in the absence of any changes in milk yield, suggesting at least in some instances alterations in milk yield and mammary blood flow could be dissociated (Mepham et al., 1984).

ALTERATIONS IN MAMMARY FUNCTION

A further feature of Bauman and Currie's 'homeorhetic signal' discussed above, is its ability to enhance the capacity of the gland to synthesize milk. An alteration in mammary function during somatotropin treatment is indicated by increases in mammary arteriovenous differences for glucose and acetate (Fullerton et al., 1989; Heap et al., 1989), NEFA's (McDowell et al., 1987b) and triglycerides (Davis et al., 1988b). However there are substantial differences between studies with respect to which metabolites show increased mammary extraction. These differences probably relate to differing energy states of the animals or to techniques of sampling mammary vein blood. This latter point is crucial for correct interpretation of the results, since unless the external pudic vein is occluded to ensure that total mammary venous drainage is obtained via the milk vein, uncontaminated by venous blood draining non-mammary tissues, then substantial errors may be experienced (Heap et al., 1989).

It may be felt that there are limitations to the value of the data obtained from these types of measurement in demonstrating enhanced mammary function in vivo because of the heterogenous nature of the glandular tissues. However, the mammary gland of lactating animals consists largely of secretory epithelium, so the relative contribution to total uptake by non-secretory tissue is probably very small and unlikely to incur serious errors.

Evidence for an additional effect of somatotropin on mammary function is provided by a recent study of the influence of recombinant bovine somatotropin on milk yield and mammary blood flow in Jersey cows in late lactation (Heap et al., 1989). Before treatment the mammary blood flow to milk yield ratio was 700:1, a value typical of this stage of lactation (Linzell, 1974). After 7 days of treatment with somatotropin this ratio decreased to 415:1, a value characteristic of cows at peak lactation (Linzell, 1974). These data indicate that the galactopoietic effect produced by recombinant somatotropin was associated not only with an increase in mammary blood flow but also, for example, with enhanced mammary extraction of certain substrates.

MEDIATION OF EFFECTS BY IGF-I

As stated previously, the absence of functional receptors for somatotropin in lactating mammary tissue indicates that it is unlikely to have a direct effect on mammary secretory function. This claim is consistent with the lack of effect of somatotropin on mammary tissue in vitro (Skarda et al., 1982; Gertler, Cohen & Maoz, 1983; Goodman et al., 1983) and the lack of stimulation of milk secretion by somatotropin infused directly into the arterial blood supplying the mammary gland (McDowell, Hart and Kirby, 1987). According to the hypothesis of Salmon and Daughaday (1957), somatotropin action may be mediated by somatomedins, specifically IGF-I. The demonstration in vitro of direct effects of IGF-I on mammary growth (Imagawa et al., 1986; Winder & Forsyth, 1986; Ethier, Kudla and Cundiff, 1987) and on development of the glucose transport system of mammary epithelial cells (Prosser et al., 1987c) suggest that somatotropin action in the mammary gland may also be mediated by IGF-I. In keeping with this, plasma concentrations of IGF-I are increased 3 to 4-fold during short-term administration of somatotropin to lactating cows (Davis et al., 1987; Prosser, Fleet and Corps, 1988), which, when account is taken of the increased mammary blood flow, results in a 5-fold increase in the calculated amount of IGF-I perfusing the gland.

However, the difficulty in interpreting such data is that circulating IGF-I is largely bound to specific binding proteins (Ooi and Herington, 1988). The major binding protein species has a molecular weight of 150 kDa while some IGF-I is associated with a 50 kDa protein. During somatotropin treatment of lactating cows the binding activity of the 150 kDa protein species increases in parallel with the concentrations of IGF-I (Prosser, Fleet and Corps, 1988). The association of IGF-I with these proteins restricts its movement out of the vascular space and attenuates its biological activity (Meulig, Zapf and Froesch, 1978; Zapf et al., 1986).

IGF-I may not act exclusively as an endocrine factor but also as an autocrine and/or paracrine factor, so that plasma levels may reflect cummulative production by different tissues. We have found that levels of immunoreactive IGF-I in mammary tissue of lactating goats treated for 4 days with pituitary-derived somatotropin were 3-fold higher than those of untreated animals (Prosser, Fleet and Heap, 1989). The tissue was taken 24-30 hours after the last injection of somatotropin, when circulating somatotropin had returned to basal values but plasma IGF-I was still elevated 2-fold. Milk yield was increased by an average of 15 $\pm$ 4% in treated animals.

The galactopoietic effect of somatotropin is accompanied by an increased secretion of IGF-I into milk (Prosser et al., 1987a; Prosser, Fleet and Corps, 1988), although peak levels attained were within the normal physiological range of dairy ruminants (Malvern et al., 1987) and lactating women (Corps et al., 1988). In cows, changes in milk concentrations of IGF-I slightly preceded any changes in milk secretion rate (Prosser, Fleet and Corps, 1988). In goats, four animals treated with somatotropin exhibited an increase in milk yield whereas one animal did not (Prosser et al., 1987a). Plasma levels of IGF-I were 2-3 fold higher in all five animals but IGF-I levels in milk only increased in the four responding animals.

Whether tissue and milk IGF-I is derived from the circulation or from local synthesis is not known. Minor amounts of mRNA for IGF-I have been detected in the normally lactating mammary gland of the rat (Murphy, Bell and Friesen, 1987) and there is evidence for the synthesis of immunoreactive IGF-I by lactating bovine mammary explants in culture (Campbell and Baumrucker, 1988). However, there is as yet no evidence for increased mammary production during somatotropin treatment. Studies in the lactating goat also show that IGF-I can be transferred from blood into milk so it is probable that at least some of the IGF-I in mammary tissue was derived from the circulation (Prosser et al., 1987b). The increased tissue levels of IGF-I and its augmented secretion into milk may reflect a role for IGF-I in the regulation of mammary gland function. In any case it would seem that the mammary epithelium is exposed to a greater amount of IGF-I during somatotropin treatment.

More direct evidence that IGF-I acts to enhance milk secretion in vivo is provided by a recent study of lactating goats (Prosser et al., 1988). IGF-I was infused close-arterially into the one gland at a rate of 1.1

nmol/min for 6 h. This resulted in an increase in the rate of milk secretion by the infused gland of 30 ± 5% compared with that in the non-infused gland of 15 ± 4% ($P < 0.05$). Concentrations of IGF-I in milk of the infused gland were increased 82 ± 22% compared with only 14 ± 9% ($P < 0.01$) in the milk of the non-infused gland. The rapidity of the response (the first significant increase occurred 4 h into the infusion) implies that IGF-I stimulated the activity of pre-existing cells rather than cellular proliferation. The unilateral response suggests that IGF-I acted directly on the mammary epithelium to enhance milk synthesis.

Circulating levels of IGF-I increased from 32.4 ± 0.8 nmol/l before, to a maximum of 53.6 ± 1.9 nmol/l 4 h after infusion commenced, but only 5% and 14% of the IGF-I was present in the free form 2 and 4 h after the start of infusion, respectively. Thus the delayed response in the non-infused gland in terms of milk yield and IGF-I concentrations presumably reflects the low transfer and efficacy of re-circulating protein-bound IGF-I. This contrasts with the effect of non-protein bound IGF-I in the infusate. Based on an infusion rate of 1.1 nmol/min and mammary blood flow of 378-543 ml/min, the infused gland would have been continuously exposed to an additional amount of free IGF-I at a concentration of 2-3 nmol/l.

It is notable that blood flow through the infused gland was increased 44% during the infusion of IGF-I but not during saline infusion. As stated earlier mammary blood flow is increased during somatotropin treatment of lactating cows (Davis et al., 1988a; Fullerton et al., 1989; Prosser, Fleet and Corps, 1988) and goats (Mepham et al., 1984) with somatotropin. Thus the possibility is raised that IGF-I is an important mediator of somatotropin regulation of mammary blood flow. The mechanism of this effect may involve direct stimulation of the vascular system by IGF-I or an indirect stimulation via the mammary release of local vasodilatory compounds subsequent to the enhancement of mammary metabolic activity.

In contrast to alterations in mammary function and blood flow, changes in whole body metabolism probably relate more to somatotropin acting in concert with other metabolic hormones than via IGF-I. Rat and human adipocytes do not possess type I IGF receptors (Massagué and Czech, 1982; Di Girolamo et al., 1986). Neither IGF-I or IGF-II were capable of replacing somatotropin in the suppression of glucose oxidation in 3T3-F442A adipocytes in culture (Schwartz, Foster and Satin, 1985). Moreover, in vitro studies, using ovine (Vernon & Finley, 1988) and bovine (Etherton and Evock, 1986) adipose tissue, showed that IGF-I only stimulated lipogenesis and glucose oxidation at concentrations high enough for IGF-I to bind to insulin receptors. The presence of IGF binding proteins in plasma would preclude any effect of circulating IGF-I on adipocyte function in vivo.

In contrast to the claims of Etherton and Evock (1986), Lewis et al. (1988) suggested that IGF-I and IGF-II are both lipolytic at low concentrations in adipose tissue from lambs. As IGF-II appeared more potent than IGF-I, the authors suggested that the effect of IGF-I was mediated via the type II receptor. However, it is difficult to determine the physiological significance of these results in view of studies in rats showing that somatotropin decreases the number of type II receptors on plasma membrane of adipocytes (Wardzala et al., 1984; Lönnroth et al., 1987).

Previously it was thought unlikely that somatotropin acted to enhance lipolysis directly (Davidson, 1987), as highly purified pituitary and recombinant somatotropin failed to elicit the same response in isolated adipocytes (see Vernon and Flint, 1989). This may be partly related to the fact that the onset of these effects require several days exposure to somatotropin and primary cultures of adipocytes are usually not viable for extended periods of time. In addition, a more important effect of somatotropin is its ability to alter the responsiveness of adipocytes to other lipolytic and lipogenic factors e.g. catecholamines and insulin (Vernon and Finley, 1988). Thus the metabolic effects elicited by somatotropin may be the result of direct action on adipocytes, for instance, but manifested only in the presence of the complete endocrine environment. For a more detailed discussion of the mechanism of action of somatotropin on nutrient mobilization the reader is referred to the paper by Vernon, R.G. in this volume.

CONCLUDING REMARKS

Somatotropin treatment of lactating ruminants to enhance milk yield is the first proposed large-scale application of recombinant technology for the manipulation of performance of domestic animals. It is evident from this review that much research has been devoted to understanding the nature of the mechanisms involved. The result of this effort demonstrates the multiplicity of events that are associated with the galactopoietic effect as well as the considerable variation in the manner in which the same gross increase in milk yield is achieved. The changes may be classified into three major areas. The primary effect appears to be altered nutrient utilization and mobilization in non-mammary tissues, which 'spares' essential nutrients for milk synthesis. This is achieved by direct effects of somatotropin on liver and adipose tissue function, for instance, but also via alterations in the responsiveness of the tissues to other metabolic hormones. The second effect is a preferential increase in blood flow to the mammary gland, although it is not clear whether this is a cause or consequence of the increase in milk secretion. The third effect of somatotropin is to alter mammary function itself. In this instance it seems likely that somatotropin acts via IGF-I to stimulate milk synthesis. Any of these events alone is capable of enhancing milk yield to some extent, but the effect would seem to be greater when they act in unison.

While these are the principal events underlying the mechanism of action of somatotropin, certain aspects remain undefined. In particular, there is a need to determine the influence of the number, affinity and tissue distribution of somatogenic receptors on the response as well as the factors regulating them during lactation. Also the relationship between cardiovascular changes and milk secretory activity, the precise nature of the alterations in mammary function and the role of blood-borne or locally synthesized IGF-I in inducing these effects require clarification. Further detailed investigations into these areas should offer the means to manage milk secretion in dairy ruminants in an even more defined manner.

ACKNOWLEDGEMENTS

We wish to thank Dr R B Heap for his comments and Mrs J Hood for typing the manuscript. The work was supported by an AFRC Link Grant.

REFERENCES

Akers, R.M. 1985. Lactogenic hormones: binding sites, mammary growth, secretory cell differentiation and milk biosynthesis in ruminants. J. Dairy Sci. 68, 501-509.

Akers, R.M. and Keys, J.E. 1984. Characterization of lactogenic hormone binding to membranes from ovine and bovine mammary gland and liver. J. Dairy Sci. 67, 2224-2235.

Asimov, G.J. and Krouze, N.K. 1937. The lactogenic preparations from the anterior pituitary and the increase of milk yield in cows. J. Dairy Sci. 20, 289-306.

Bauman, D.E. and Currie, W.B. 1980. Partitioning of nutrients during pregnancy and lactation: a review of mechanisms involving homeostasis and homeorhesis. J. Dairy Sci. 63, 1514-1529.

Bauman, D.E., Eppard, P.J., De Geeter, M.J. and Lanza, G.M. 1985. Responses of high producing dairy cows to long-term treatment with pituitary somatotropin and recombinant somatotropin. J. Dairy Sci. 68, 1352-1362.

Baxter, R.C., Zaltsman, Z. and Turtle, J.R. 1984. Rat growth hormone (GH) but not prolactin (PRL) induces both GH and PRL receptors in female rat liver. Endocr. 114, 1893-1901.

Beckers, J.F. 1987. Coexistence of lactogenic and somatogenic receptors in bovine mammary gland. Proceedings of the 1st European Congress of Endocrinology, June 21-25. p. 138.

Bines, J.A., Hart, I.C. and Morant, S.V. 1980. Endocrine control of energy metabolism in the cow: the effect on milk yield and levels of some blood constituents of injecting growth hormone and growth hormone fragments. Br. J. Nutr. 43, 179-188.

Bitman, J., Wood, D.L., Tyrrell, H.F., Bauman, D.E., Peel, C.J., Brown, A.C.G. and Reynolds, P.J. 1984. Blood and milk lipid responses induced by growth hormone administration in lactating cows. J. Dairy Sci. 67, 2873-2880.

Boyd, R.D. and Bauman, D.E. 1988. Mechanism of action for somatotropin in growth. In "Current Concepts of Animal Growth Regulation". (Eds D.R. Campion, G.J. Hausman and R.J. Martin). (Plenum, N.Y.).

Breier, B.H., Gluckman, P.D. and Bass, J.J. 1988. The somatotrophic axis in young steers: influence of nutritional status and oestradiol-17β on hepatic high and low-affinity somatotrophic binding sites. J. Endocr. 116, 169-177.

Campbell, P.G. and Baumrucker, C.R. 1988. Secretion of immunoreactive insulin-like growth factor I and its binding protein from the bovine mammary gland in vitro. Proc. Endocr. Soc. 70th Annual Meeting, June 8-11, Abst. 510.

Corps, A.N., Brown, K.D., Rees, L.H., Carr, J. and Prosser, C.G. 1988. The insulin-like growth factor I content in human milk increases between early and full lactation. J. Clin. Endocr. Metab. 67, 25-29.

Davidson, M.B. 1987. Effect of growth hormone on carbohydrate and lipid metabolism. Endocr. Rev. 8, 115-131.

Davis, S.R., Gluckman, P.D., Hart, I.C. and Henderson, H.V. 1987. Effects of injecting growth hormone or thyroxine on milk production and blood plasma concentrations of insulin-like growth factors I and II in dairy cows. J. Endocr. 114, 17-24.

Davis, S.R., Collier, R.J., McNamara, J.P., Head, H.H. and Sussman, W. 1988a. Effects of thyroxine and growth hormone treatment of dairy cows on milk yield. J. Anim. Sci. 66, 70-79.

Davis, S.R., Collier, R.J., McNamara, J.P., Head, H.H., Croon, W.J. and Wilcox, C.J. 1988b. Effects of thyroxine and growth hormone treatment of dairy cows on mammary uptake of glucose, oxygen and other milk fat precursors. J. Anim. Sci. 66, 80-89.

Di Girolamo, M., Edén, S., Enberg, G., Isaksson, O., Lönron, P., Hall, K. and Smith, U. 1986. Specific binding of human growth hormone but not insulin-like growth factors by human adipocytes. FEBS Lett. 205, 15-19.

Elvinger, F., Head, H.H., Wilcox, C.J., Natzke, R.P. and Eggert, R.G. 1988. Effects of administration of bovine somatotropin on milk yield and composition. J. Dairy Sci. 71, 1515-1525.

Eppard, P.J., Bauman, D.E. and McCutcheon, S.N. 1985. Effect of dose of bovine growth hormone on lactation of dairy cows. J. Dairy Sci. 68, 1109-1115.

Eppard, P.J., Bauman, D.E., Bitman, J., Wood, D., Akers, R.M. and House, W.A. 1985. Effect of dose of bovine growth hormone on milk composition: α-lactalbumin, fatty acids and mineral elements. J. Dairy Sci. 68, 3047-3054

Etherton, T.D. 1989. The mechanisms by which porcine growth hormone improves pig growth performance. In "Biotechnology in Growth Regulation" (Eds R.B. Heap, C.G. Prosser and G.E. Lamming). (Butterowrth, London).

Etherton, T.D. and Evock, C.M. 1986. Stimulation of lipogenesis in bovine adipose tissue by insulin and insulin-like growth factor. J. Anim. Sci. 62, 357-362.

Ethier, S.P., Kudla, A. and Cundiff, K.C. 1987. Influence of hormone and growth factor interactions on the proliferative potential of normal rat mammary epithelial cells in vitro. J. Cell Physiol. 132, 161-167.

Evans, H.M. and Simpson, M.E. 1931. Hormones of the anterior hypophysis. Am. J. Physiol. 98, 511-546.

Fleet, I.R., Fullerton, F.M., Heap, R.B., Mepham, T.B., Gluckman, P.D. and Hart, I.C. 1988. Cardiovascular and metabolic responses during growth hormone treatment of lactating sheep. J. Dairy Res. 55, 479-485.

Fullerton, F.M., Fleet, I.R., Heap, R.B., Hart,I.C. and Mepham, T.B. 1989. Cardiovascular responses and mammary substrate uptake in Jersey cows treated with pituitary-derived growth hormone during late lactation. J. Dairy Res. (in press).

Gertler, A., Cohen, N. and Maoz, A. 1983. Human growth hormone but not ovine or bovine growth hormone exhibit galactopoietic prolactin-like activity in organ culture from bovine lactating mammary gland. Molec. Cell Endocr. 33, 169-182.

Gluckman, P.D. and Breier, B.H. 1989. The regulation of the growth hormone receptor. In "Biotechnology in Growth Regulation" (Eds R.B. Heap, C.G. Prosser and G.E. Lamming). (Butterworth, London).

Gluckman, P.D. Butler, J.H. and Elliot, T.B. 1983. The ontogeny of somatotrophic binding sites in ovine hepatic membranes. Endocr. 112, 1607-1612.

Goodman, G.T., Akers, R.M., Friderici, K.H. and Tucker, H.A. 1983. Hormonal regulation of α-lactalbumin secretion from bovine mammary tissue cultured in vitro. Endocr. 112, 1324-1330.

Hanwell, A. and Linzell, J.L. 1973. The time course of cardiovascular changes in lactation in the rat. J. Physiol. 233, 93-109.

Heap, R.B., Fleet, I.R., Fullerton, F.M., Davis, A.J., Goode, J.A., Hart, I.C., Pendleton, J.W., Prosser, C.G., Sylvester, L.M. and Mepham, T.B. 1989. A comparison of the mechanisms of action of bovine pituitary-derived and recombinant somatotropin (ST) in inducing galactopoiesis in the cow during late lactation. In "Biotechnology in Growth Regulation". (Eds R.B. Heap, C.G. Prosser and G.E. Lamming). (Butterworths, London).

Hughes, J.P. 1979. Identification and characterization of high and low affinity binding sites for growth hormone in rabbit liver. Endocr. 105, 414-420.

Imagawa, W., Spencer, E.M., Larson, L. and Nandi, S. 1986. Somatomedin C substitutes for insulin for the growth of mammary epithelial cells from normal virgin mice in serum-free collagen gel cell culture. Endocr. 119, 2695-2699.

Isaksson, O.G., Lindahl, A., Nilsson, A. and Isgaard, J. 1987. Mechanism of the stimulatory effect of growth hormone on longitudinal growth hormone. Endocr. Rev. 8, 426-438.

Lanza, G.M., Eppard, P.J., Miller, M.A., Franson, S.E., Ganguli, S., Hintz, R.L., Hammond, B.G,., Busser, S.C., Leak, R.K. and Metzger, L.E. 1988. Response of lactating dairy cows to multiple injections of sometribove, USAN (recombinant methionyl bovine somatotropin) in a prolonged release system. Part III Changes in circulating analytes. J. Dairy Sci. 71, (Suppl. 1) 184.

Lee, M.O. and Schaffer, N.K. 1934. Anterior pituitary growth hormone and the composition of growth. J. Nutr. 7, 377.

Lesniak, M.A. and Roth, J. 1976. Regulation of receptor concentration by homologous hormone. Effect of human growth hormone on its receptor in IM-9 lymphocytes. J. Biol. Chem. 251, 3720-3729.

Lewis, K.J., Molan, P.C., Bass, J.J. and Gluckman, P.D. 1988. The lipolytic activity of low concentrations of insulin-like growth factors in ovine adipose tissue. Endocr. 122, 2554-2557.

Linzell, J.L. 1974. Mammary blood flow and methods of identifying and measuring precursors of milk. In "Lactation" (Eds B.L. Larson and V.R. Smith). (Academic Press, N.Y.) vol. 1, pp. 143-225.

Lönnroth, P., Assmundsson, K., Edén, S., Enberg, G., Gouse, I., Hall, K.

and Smith, U. 1987. Regulation of insulin-like growth factor II receptors by growth hormone and insulin in rat adipocytes. Proc. Natl. Acad. Sci., U.S.A. 84, 3619-3622.

Lynch, J.M., Barbano, D.M., Bauman, D.E. and Hartnell, G.F. 1988. Influence of sometribove (recombinant methionyl bovine somatotropin) on the protein and fatty acid composition of milk. J. Dairy Sci. 71, (Suppl. 1) 100.

Maes, M., Underwood, L.E., Gerard, G. and Ketelslegers, J.-M. 1984. Relationship between plasma somatomedin-C and liver somatogenic binding site in neonatal rats during malnutrition and after short and long term refeeding. Endocr. 115, 786-792.

Maiter, D., Maes, M., Underwood, L.E., Fliesen, T., Gerard, G. and Ketelslegers, J.-M. 1988. Early changes in serum concentrations of somatomedin-C induced by dietary protein deprivation in rats: contributions of growth hormone receptor and post-receptor defects. J. Endocr. 118, 113-120.

Malvern, P.V., Head, H.H., Collier, R.J. and Buonomo, F.C. (1987). Periparturient changes in secretion and mammary uptake of insulin and in concentrations of insulin and insulin-like growth factors in milk of dairy cows. J. Dairy Sci. 70, 2254-2265.

Massagué, J. and Czech, M.P. 1982. The subunit structures of two distinct receptors for insulin-like growth factors I and II and their relationship to the insulin receptor. J. Biol. Chem. 257, 5038-5045.

McDowell, G.H., Hart, I.C. and Kirby, A.C. 1987. Local intra-arterial infusion of growth hormone into the mammary glands of sheep and goats: effects on milk yield and composition, plasma hormones and metabolites. Aust. J. Biol. Sci. 40, 181-189.

McDowell, G.H., Hart, I.C., Bines, J.A., Lindsay, D.B. and Kirby, A.C. 1987a. Effects of pituitary-derived bovine growth hormone on production parameters and biokinetics of key metabolites in lactating dairy cows at peak and mid-lactation. Aust. J. Biol. Sci. 40, 191-202.

McDowell, G.H., Gooden, J.M., Leenanuruska, M.J. and English, A.W. 1987b. Effects of exogenous growth hormone on milk production and nutrient uptake by muscle and mammary tissues of dairy cows in mid-lactation. Aust. J. Biol. Sci. 40, 295-306.

Mepham, T.B., Lawrence, S.E., Peters, A.R. and Hart, I.C. 1984. Effects of growth hormone on mammary function in lactating goats. Horm. Metab. Res. 16, 148-253.

Merimee, T.J., Zapf, J. and Froesch, E.R. 1982. Insulin-like growth factors in the fed and fasted states. J. Clin. Endocr. Metabol. 55, 999-1002.

Meulig, C., Zapf, J. and Froesch, E.R. 1978. NSILA-carrier protein abolishes the action of non-suppressible insulin-like activity (NSILA-S) on perfused rat heart. Diabetologia 14, 253-259.

Murphy, L.J., Bell, G.I. and Friesen, H.G. 1987. Tissue distribution of insulin-like growth factor I and II messenger-ribonucleic acid in the adult rat. Endocr. 120, 1279-1282.

Murphy, L.J., Vrhovsek, E. and Lazarus, L. 1983. Identification and characterization of specific growth hormone receptors in cultured human fibroblasts. J. Clin. Endocr. Metab. 57, 1117-1124.

Ooi, G.T. and Herington, A.C. 1988. The biological and structural characterization of specific serum binding proteins for the insulin-like growth factors. J. Endocr. 118, 7-18.

Nissley, S.P. and Rechler, M.W. 1985. Insulin-like growth factors: biosynthesis, receptors and carrier proteins. Hormonal Proteins and Peptides. 12, 128-203.

Peel, C.J. and Bauman, D.E. 1987. Somatotropin and lactation. J. Dairy Sci. 70, 474-486.

Peel, C.J., Eppard, P.J. and Hard, D.L. 1989. Evaluation of sometribove (methionyl bovine somatotropin) in toxicology and clinical trials in Europe and the United States. In "Biotechnology in Growth Regulation". (Eds R.B. Heap, C.G. Prosser and G.E. Lamming). (Butterworth, London).

Peel, C.J., Fronk, T.J., Bauman, D.E. and Gorewit, R.C. 1983. Effect of exogenous growth hormone in early and late lactation on lactational performance in dairy cows. J. Dairy Sci. 66, 776-782.

Pocius, P.A. and Herbein, J.H. 1986. Effects of in vivo administration of growth hormone on milk production and in vitro hepatic metabolism in dairy cattle. J. Dairy Sci. 69, 713-720.

Prosser, C.G., Fleet, I.R. and Corps, A.N. 1988. Increased secretion of insulin-like growth factor I into milk of cows treated with recombinantly derived bovine growth hormone. J. Dairy Res. (in press).

Prosser, C.G., Fleet, I.R. and Heap, R.B. 1989. Action of IGF-1 on mammary function. In "Biotechnology in Growth Regulation". (Eds R.B. Heap, C.G. Prosser and G.E. Lamming). (Butterworths, London).

Prosser, C.G., Fleet, I.R., Hart, I.C. and Heap, R.B. 1987a. Changes in concentrations of insulin-like growth factor I (IGF-I) in milk during bovine growth hormone treatment in the goat. J. Endocr. 112 (Suppl.), 65.

Prosser, C.G., Davis, A.J., Fleet, I.R., Rees, L.H. and Heap, R.B. 1987b. Mechanism of transfer of IGF-I into milk. J. Endocr. 115 (Suppl.), 91.

Prosser. C.G., Sankaran, L., Hennighausen, L. and Topper, Y.J. 1987c. Comparison of the roles of insulin and insulin-like growth factor I in casein gene expression and in the development of α-lactalbumin and glucose transport activities in the mouse mammary epithelial cell. Endocr. 120, 1411-1416.

Prosser, C.G., Fleet, I.R., Corps, A.N., Heap, R.B. and Froesch, E.R. 1988. Increased milk secretion and mammary blood flow during close-arterial infusion of insulin-like growth factor I (IGF-I) into the mammary gland of the goat. J. Endocr. 117 (Suppl.), 248.

Ronge, H., Blum, J., Clement, C., Jans, F., Leuenberger, H. and Binder, H. 1988. Somatomedin C in dairy cows related to energy and protein supply and to milk production. Anim. Prod. 47, 165-183.

Salmon, W.D. and Daughaday, W.H. 1957. A hormonally controlled serum factor which stimulates sulphate incorporation by cartilage in vitro. J. Lab. Clin. Med. 49, 825-829.

Schatz, H., Katsilambros, N., Hinz, M., Voigt, K.H., Nierle, C. and Pfeiffer, E.F. 1973. Hypophysis and function of pancreatic islets. II. The effect of substitution with growth hormone and corticotrophin on insulin secretion and biosynthesis of proinsulin and insulin in isolated pancreatic islets of hypophysectomized rats. Diabetologia 9, 140-144.

Schemm, S.R. and Deaver, D.R. 1988. Administration of recombinant bovine somatotropin (rbst) to lactating dairy cows beginning at 35 and 70 days postpartum. IV. Effect on pituitary function and plasma estradiol. J. Dairy Sci. 71, (Suppl. 1) 167.

Schwartz, J., Foster, C.M. and Satin, M.S. 1985. Growth hormone and insulin-like growth factors I and II produce distinct alterations in glucose metabolism in 3T3-F442A adipocytes. Proc. Natl. Acad. Sci., U.S.A. 82, 8724-8728.

Skarda, J., Urbanova, E., Becka, S., Houdebine, L.M., Delouis, C., Pichova, D., Picha, J. and Pilek, J. 1982. Effect of bovine growth hormone on development of goat mammary tissue in organ culture. Endocr. Expt. 16, 19-31.

Soderholm, C.G., Otterby, D.E., Linn, J.G., Ehle, F.R., Wheaton, J.E.,

Hansen, W.P. and Annexstad, R.J. 1988. Effects of recombinant bovine somatotropin on milk production, body composition and physiological parameters. J. Dairy Sci. 71, 355-365.

Tsushima, T. and Friesen, H.G. 1973. Radioreceptor assay for growth hormone. J. Clin. Endocr. Metab. 37, 334-337.

Vernon, R.G. and Finley, E. 1988. Roles of insulin and growth hormone in the adaptations of fatty acid synthesis in white adipose tissue during the lactation cycle in sheep. Biochem. J. 256, 873-878.

Vernon, R.G. and Flint, D.J. 1989. Role of growth hormone in the regulation of adipocyte growth and function. In "Biotechnology in Growth Regulation" (Eds R.B. Heap, C.G. Prosser and G.E. Lamming). (Butterworth, London).

Wallis, M. 1989. Species specificity and structure-function relationships of growth hormone. In "Biotechnology in Growth Regulation" (Eds R.B. Heap, C.G. Prosser and G.E. Lamming). (Butterworth, London).

Wardzala, L.J., Simpson, I.A., Rechler, M.M. and Cushman, S.W. 1984. Potential mechanism of the stimulatory action of insulin on insulin like growth factor II binding to the isolated rat adipose cell. J. Biol. Chem. 259, 8378-8383.

Winder, S.J. and Forsyth, I.A. 1986. Insulin-like growth factor I (IGF-I) is a potent mitogen for ovine mammary epithelial cells. J. Endocr. 106 (Suppl.), 141.

Young, F.G. 1947. Experimental stimulation (galactopoiesis) of lactation. Brit. Med. Bull. 5, 155-161.

Zapf, J., Hauri, Ch., Waldvogel, M. and Froesch, E.R. 1986. Acute metabolic effects and half-lives of intravenously administered insulin-like growth factors I and II in normal and hypophysectomized rats. J. Clin. Invest. 77, 1768-1775.

VARIATION OF BST AND IGF-I CONCENTRATIONS IN BLOOD PLASMA OF CATTLE

D. Schams, U. Winkler, M. Theyerl-Abele, A. Prokopp

Institut für Physiologie
Südd. Versuchs- und Forschungsanstalt für Milchwirtschaft
Technische Universität München, 8050 Freising-Weihenstephan
FRG

ABSTRACT

Secretion of bST and IGF-I was measured in growing male and female cattle, during the lactation period and during treatment of cows with a slow release preparation of bST. In general, males exhibit higher levels of bST and IGF-I than females. BST concentrations are higher in the first month of age, are lower thereafter and increase again to reach a maximum around puberty. This is mainly due to an augmentation of amplitude but not basal secretion of bST. IGF-I is low in the first month, reveals an initial increase up to the 3rd -5th month, and a 2nd one approaching puberty. During lactation concentrations of bST increased during peak lactation, decreased thereafter and reached a lower plateau about 18 weeks pp and remained at that level for the rest of the lactation. The increase of bST pp is related to an enhanced episodic secretory activity. In general bST basal levels and absolute values of amplitudes are lower in cows than in growing cattle. Treatment of cows with a sustained release preparation of bST increased bST and IGF-I levels 2-5 fold. For bST there is still episodic activity measurable. The main difference is an increase of basal levels to a higher plateau in treated animals. Absolute concentrations are still within the physiological range of growing animals. After termination of treatment bST and IGF-I concentrations resume levels of control animals within a short period of time.

INTRODUCTION

Hormonal regulation of growth and lactation is extremely complex involving many different hormones, nutrient availability and multiple interactions between the endocrine system and genetic and environmental factors. Somatotropin (bST) and its related growth factor (insulin-like growth factor, IGF-I) are believed to play an essential role in this concert. Pituitary and recombinant-derived bovine somatotropin increase milk production and growth of veal calves (for review see Karg 1987, Karg &

Mayer 1987). In connection with the use of rbST different aspects of animal and human health protection are discussed. For the estimation of a potential risc the variation of hormone levels under physiological conditions during growth and lactation and during treatment with rbST is an important prerequisite for proper evaluation.

MATERIAL AND METHODS

Experiment 1

Blood levels of STH and IGF-I during growth from birth until puberty (Schams et al. 1988): Concentrations of somatotropin and insulin-like growth factor I were measured in the blood plasma of 4 male and 2 female cattle of the Brown Swiss breed. The animals were fed milk replacer for 12 weeks and hay, corn silage and concentrate pellets afterwards. Until age of 6 months blood was collected in the morning and evening via a permanent canula inserted into the Vena cava caudalis. Additionally frequent sampling periods (6h at 15 min intervals) were included. After age of 6 months blood was collected 3 times/week per needle puncture of the jugular vein or by a permanent canula (24 h at 30 min intervals) inserted into the jugular vein at 4 week intervals.

Experiment 2

Effect of sex on concentrations of bST and IGF-I: This experiment was done in cooperation with the Animal Nutrition Department in Weihenstephan. Heifers, bulls and steers (12/group of the Fleckvieh breed) were fed ad lib under defined nutritional conditions. Blood was collected by needle puncture of the jugular vein at two week intervals from 5 month of age until an age of 15-16 months.

Experiment 3

BST and IGF-I were analysed during lactation of Brown Swiss cows. Blood was collected from the jugular vein by needle puncture or by frequent bleedings (12h, 30 min intervals) at certain intervals throughout lactation.

Experiment 4

Application of rbST in Fleckvieh cows: This experiment was performed on 12 cows in a change-over design. For detailes see Graf et al. (1988). Treatment of a slow release device rbST preparation (500 mg s.c. at 2 week intervals) started after 9 weeks pp for 10 weeks, followed by a 4 week non-treatment period and again treatment of the previous control group for 10 weeks. 5 time courses with frequent bleedings (7 days after the injection, 24 h at 30 min intervals) were performed.

Experiment 5.

Treatment of Fleckvieh cows with rbST. This experiment was performed with 8 cows (4 treated, 4 controls). A slow release preparation (640 mg s.c.) was given at 4 week intervals. Blood was collected by needle puncture of the jugular vein on days -1, 1, 3, 6, 8, 10, 13, 15, 17, 20, 22, 24, 27 after the third injection and on day 1, 3, 6, 8, 10 and 13 after the fourth injection.

Hormone analysis

Somatotropin and IGF-I were evaluated as described by Schams in this volume.

Blood collection

Blood (10 ml) was collected in tubes containing 200 µl of a solution with 0.3 M EDTA and 1 % aspirin. Blood was cooled in ice water for 5 min and centrifuged at $4^\circ C$ for 20 min at 3000 x g. Plasma was stored at $-20^\circ C$ until analysis.

RESULTS

BST and IGF-I during growth

The evaluation of blood samples of exp. 1 for somatotropin is shown in table 1 and for IGF-I in Fig. 1.

The increase of bST is mainly due to an increase of the amplitude but not of basal secretion for bST. In females bST levels are decreasing much earlier than in bulls. During the period with high amplitudes individual peak values may be higher than 100 ng/ml. Typical examples for the episodic secretion pattern are shown in Figure 2. In general, males exhibit higher levels of bST and IGF-I than females.

Table 1. Somatotropin levels in growing Brown Swiss cattle (mean; ng/ml, evaluation by PULSAR program)

Age in month	Overall mean male	female	Basal concentr. male	female	Amplitude male	female	Frequency of episodes/24h male	female
1	14.7	11.6	10.6	8.8	20.5	15.0	9.1	8.7
2	4.4	4.1	3.7	2.9	7.6	5.8	10.0	10.0
3	7.8	5.0	5.2	3.7	11.1	4.6	10.3	9.0
4	11.2	3.9	6.8	3.1	16.7	5.7	9.4	8.7
5	22.4	3.3	12.2	2.8	35.1	3.9	8.3	6.7
6	22.8	8.7	15.8	5.2	37.5	10.7	5.2	14.5
7	15.2	19.0	7.0	6.2	29.3	48.3	8.6	6.3
8	19.9	11.1	7.3	6.4	34.5	16.6	9.3	6.5
9	26.4	8.7	8.3	4.8	48.9	12.0	8.3	7.0
10	20.5	6.9	8.3	4.4	33.4	8.4	7.2	5.3
11	17.8	7.1	7.0	4.6	29.1	9.7	7.8	7.0
12	15.1	6.3	7.5	4.9	23.5	5.3	6.5	6.0
13	12.6	7.1	5.1	5.5	17.8	7.3	8.3	4.5
14		6.2		5.7		4.3		2.0

This can be also seen from exp. 2. The overall mean for bST and IGF-I obtained from bulls, steers and heifers is summarized in table 2. In this experiment again individual samples exceeded 100 ng/ml blood plasma. The main increase of IGF-I occurs around the time of puberty. Concentrations may exceed 2000 ng/ml especially in bulls.

BST and IGF-I during lactation

Figure 3 shows mean values for bST and IGF-I of 9 Brown Swiss cows from ante partum and during the lactation period. Concentrations of bST increased after parturition for about 8 weeks, decreased thereafter, reached a lower plateau 18 weeks post-partum (pp) and remained at that level for the rest of the lactation. The increase of bST pp was related to an enhanced episodic secretory activity of

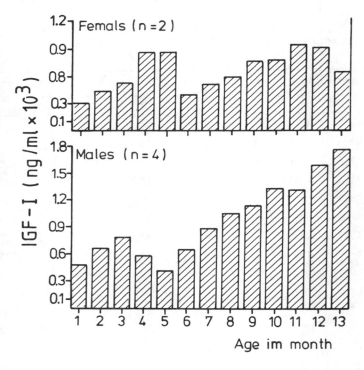

Fig. 1 Concentrations of IGF-I (mean) in growing Brown Swiss cattle (n=4 males; n = 2 females).

bST with increased basal concentrations and amplitudes. In general peak concentrations of amplitues are below 35 ng/ml. But in some cows values are measured up to 64 ng/ml. After peak lactation bST levels decreased due to a decrease of basal values and maximum of amplitudes. Frequency of amplitudes did not change significantly. During mid or late lactation occasionally some higher amplitudes (range 15-30 ng/ml) are measured. Levels of IGF-I decreased after parturition, remained for the first five weeks at the lower level and increased afterwards continously and reached a plateau about 28 week pp. In general compared with growing animals levels of bST and IGF-I are lower in cows. A comparison of bST concentrations in cows of different breeds (Brown Swiss, Fleckvieh and Holstein Friesian) indicated similar levels within the dual purpose breeds and higher concentrations during peak lactation in Holstein Friesian.

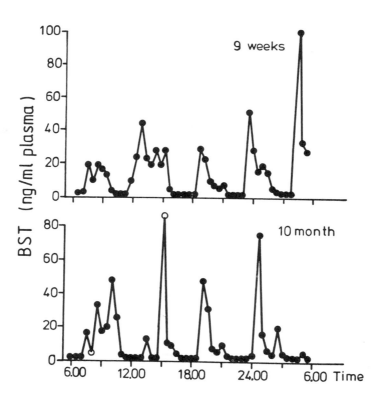

Fig. 2 Episodic secretion of bST in a 9 week and a 10 month old Fleckvieh bull

Table 2. Average concentrations of bST and IGF-I in blood of Fleckvieh cattle (n = 12/group) during growth from 5-16 month of age (mean ± S.D.) fed ad lib.

	Bulls	Steers	Heifers
BST (ng/ml)	25 ± 6.1	13.9 ± 4.4	10.1 ± 4
IGF-I (ng/ml)	1385±408	1323 ± 396	1065 ± 361

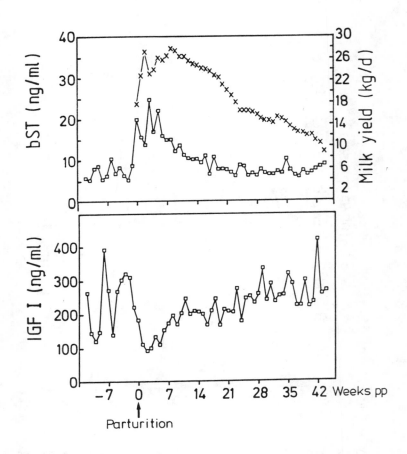

Fig. 3 Concentration of bST and IGF-I (mean) during the lactation period of Brown Swiss cows (n = 9)

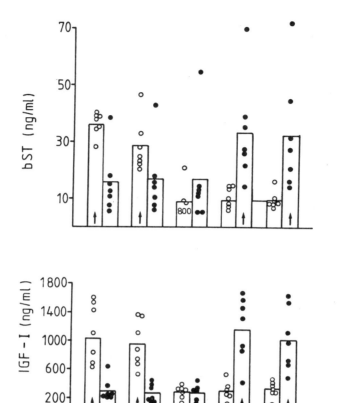

Fig. 4 Concentrations of bST and IGF-I in blood plasma of Fleckvieh cows during treatment with a slow release preparation (500 mg s.c., 2 week intervals). Each point (O group A; ● group B) represents the mean of 49 samples for one individual animal collected 7 days after injection. Treated cows are marked by an error, column represents mean for the whole group.

BST and IGF-I during treatment with bST.

Mean values are given from each time course (49 samples) for each individual animal and per group with and without treatment with rbST from experiment 4 in Figure 4.. There is a significant 2-3 fold increase for bST during treatment with a high variability between animals. Concentrations of IGF-I increased about 4-fold on average again with a high variation between cows. With one exception levels of untreated cows did not exceed 500 ng/ml whereas treated animals exceeded more or less this threshold. The variations of values within one individual animal were small indicating that treatment just increased concentrations to a higher plateau. With another group of Fleckvieh cows, different management and a preparation from another company more or less similar data were obtained (experiment 5, Figure 5). Highest levels for bST are reached 3 days after the injection and declined continously afterwards. IGF-I increased and decreased with a delay of about 2 days to bST. After 10-12 days control values were reached. In a further experiment with 8 Fleckvieh cows using the same preparation by monitoring concentrations of bST and IGF-I after the first and last (6th) injection of a treatment period more or less similar results were yielded.

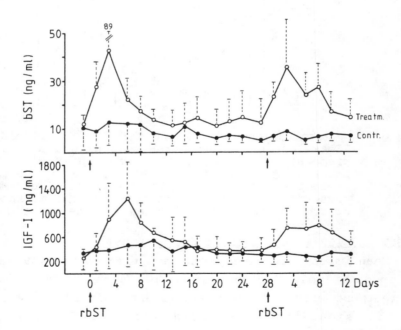

Fig. 5 Concentrations of bST and IGF-I (mean ± S.D.) in blood plasma during treatment with a slow release preparation (640 mg s.c. at 4 week intervals, third and fourth injection); 4 cows/group; ●--● controls; o--o treatment.

DISCUSSION

To our knowledge comparable continuous monitoring of bST and IGF-I concentrations during bovine growth is not available. But absolute levels of bST are comparable with reports in the literature (Purchas et al. 1970; Verde & Trenkle 1986). There is general agreement that bST levels are higher in males than in females (Irvin & Trenkle 1971; Reynaert et al. 1976). Nutrition and steroids have a modulating effect on bST and IGF-I secretion (Breier et al. 1986). Reduced feeding increased mean plasma bST concentrations and the amplitude of bST pulses. IGF-I levels decreased. The episodic nature of bST secretion agrees with reports obtained with 10 month old steers (Moseley et al. 1985; Breier et al. 1986, Wheaton et al. 1986). The increase of IGF-I in bulls is confirmed by Lund-Larsen et al. (1977) measuring IGF-I with a bioassay from 6 up to 10 month of age.

Our data obtained for bST during lactation in cows are in agreement with observations made in other breeds (mainly Holstein Friesian) by Koprowski & Tucker (1973); Smith et al. (1976); Johke & Hodate (1977); Vasilatos & Wangsness (1981); Bines & Hart (1982); Ronge & Blum (1988). Our observation that the increase of bST is mainly due to an elevation of amplitudes agrees with data from Vasilatos & Wangsness (1981). During peak lactation episodes with amplitudes up to 60 ng/ml are measured. Comparable data for IGF-I are limited. Ronge & Blum (1988) found like we a decrease of IGF-I during peak lactation parallel to the increase of bST. Absolute concentrations are lower as reported in our studies.

After treatment with exogenous bST (pituitary derived or recombinant) tremendous variations exists depending on formulation of bST, time of blood collection after the injection and dose of bST used. In most of the studies in which bST was evaluated in blood pituitary bST was injected. Mean concentrations increased after daily application of 38.5 mg, 76.9 mg from 5 to 19.5 or 30.1 ng/ml 2 h after the injection (Eppard et al. 1985). Fronk et al. (1983) reported after daily treatment with 39.6 mg/cow/d a mean increase from 5.8 ng to 48 ng/ml (single injection), to 63.7 ng/ml (injections 6 times/d) or to 25.8 ng (after infusion). A similar study (Peel et al. 1983) with 39.6 mg bST/d for 10 days caused an increase from 7.1 to 29 ng/ml (12 weeks pp) and from 6.2 to 42.4 ng/ml (35 weeks pp). With a lower dose of 22.7 mg/d for 11 days blood was collected frequently (5h, 10 min intervals/on days -9, 1 and 10 of treatment. Mean values increased from 6.9 to 14.4 (day 1) or 24.8 ng/ml (day 10) indicating a decrease in clearance rate of bST (Pocius et al. 1986). In another study by Soderholm et al. (1988) 3 different doses are given daily s.c. from 4 to 38 weeks pp (0, 10, 20.6 and 41.2 mg/d). Three hours after the last injection values increased from 4.6 to 34.5, 66.5 and 135.4 ng/ml respectively. 24 h after the injection levels were 3.8, 10.6, 15.5 and 23.0 ng/ml respectively.

I n general absolute concentrations of bST increased significantly during treatment with bST. Levels are within the range of growing animals but higher concentrations lasting longer (especially after use of a sustained release preparation) compared with relatively short lasting episodes of endogenous bST. Absolute values for IGF-I increased 2-5 fold after application of a slow release preparation in cows. But levels are still within the physiological range as observed in growing animals. After stop of treatment elevated levels decreased to control ones within a relatively short period of time. This suggests that a possible inhibition of endogenous bST is not permanent.

ACKNOWLEDGEMENTS
We thank Dr. Raiti from the National Hormone and Pituitary Program, Baltimore, Maryland, USA for the generous gift of USDA-bGH-B-1 and USDA-bGH-I-1 and Prof. P. Gluckman,

Auckland, New Zealand for the supply of IGF-I antiserum. The support by the Wilhelm Schaumann Stiftung; Elanco, Eli Lilly Comp. USA and Monsanto Comp. USA is acknowledged.

REFERENCES

Breier, B.H., Bass, J.J., Butler, J.H. and Gluckman, P. 1986. The somatotrophic axis in young steers: influence of nutritional status on pulsatile release of growth hormone and circulating concentrations of insulin-like growth factor I. J. Endocr. 111, 209-215.

Bines, J.A. and Hart, I.C. 1982. Metabolic limits to milk production, especially roles of growth hormone and insulin. J. Dairy Sci. 65, 1375-1389.

Eppard, P.J., Bauman, D.E., Bitman, J., Wood, D.L., Akers, R.M. and House, W.A. 1985. Effect of dose of bovine growth hormone on milk composition: lactalbumin, fatty acids and mineral elements. J. Dairy Sci. 68, 3047-3054.

Fronk, T.J., Peel, C., J., Bauman, D.E. and Gorewit, R.C. 1983. Comparison of different patterns of exogenous growth hormone administration on milk production in Holstein cows. J. Anim. Sci. 57, 699-705.

Graf, F., Meyer, J., Schams, D., Dennhöfer, W., Müller, G. and Kräußlich H. 1989. Effect of recombinant bovine somatotropin (BST) on physiological parameters and on milk production in German Fleckvieh cows. Zentralbl. Vet. Med., Reihe A in press

Irvin, R. and Trenkle, A. 1971. Influence of age, breed and sex on plasma hormones in cattle. J. Anim. Sci. 32, 292-295.

Johke, T. and Hodate, K. 1977. Bovine serum prolactin, growth hormone and triiodothyronine levels during late pregnancy and early lactation. Jap. J. Zootech. Sci. 48, 772-776.

Karg, H. 1987. Hormonale Manipulation des Wachstums. Übers. Tierernährung 15, 1-28.

Karg, H. and Mayer, H. 1987. Manipulation der Laktation. Übers. Tierernährung 15, 29-58.

Koprowski, J. A. and Tucker, H.A. 1973. Bovine serum growth hormone, corticoids and insulin during lactation. Endocrinology 93, 645-651.

Lund-Larsen, T.R., Sundby, A., Kruse, V. and Velle, W. 1977. Relation between growth rate, serum somatomedin and plasma testosterone in young bulls. J. Anim. Sci. 44, 189-194.

Moseley, W:M., Krabill, L.F., Friedman, A.R. and Olsen, R.F. 1985. Administration of synthetic human pancreatic growth hormone-releasing factor for five days sustains raised serum concentrations of growth hormone in steers. J. Endocr. 104, 433- 439.

Peel, C.J., Frank, T.J., Bauman, D.E. and Gorewit, R.C. 1983. Effect of growth hormone in early and late lactation on lactatonal performance of dairy cows. J.

Dairy Sci. 66, 776-782.

Pocius, P.A. and Herbein, J.H. 1986. Effects of in vivo administration of growth hormone on milk production and in vitro hepatic metabolism in dairy cattle. J. Dairy Sci. 69, 713-720.

Purchas, R.W., Macmillian, K.L. and Hafs, H.D. 1970. Pituitary and plasma growth hormone levels in bulls from birth to one year of age. J. Anim. Sci. 31, 358-363.

Reynaert, R., Marcus, S. and Peeters, P. 1976. Influences of stress, age, sex on serum growth hormone and free fatty acid levels in cattle. Horm. Metab. Res. 8, 109-114.

Ronge, H. and Blum J.W. 1988. Somatomedin C and other hormones in dairy cows around parturition, in newborn calves and in milk. J. Anim. Physiol. and Animal Nutrition 60, 168-176.

Schams, D., Winkler, U., Schallenberger, E. and Karg, H. 1988. Wachstumshormon und insulin like growth factor I (Somatomedin C) - Blutspiegel bei Rindern von der Geburt bis nach der Pubertät. Dtsch. Tierärztl. Wschr. 95, 353-408.

Smith, R.D., Hansel, W. and Coppock, C.E. 1976. Plasma growth hormone and insulin during early lactation in cows fed silage based diets. J. Dairy Sci. 59, 248-254.

Soderholm, C.G., Otterby, D.E., Linn, J.G., Ehle, F.R., Wheaton, J.E., Hansen, W.P. and Annexstad, R.J. 1988. Effects of recombinant bovine somatotropin on milk production, body composition and physiological parameters. J. Dairy Sci. 71, 355-365.

Vasilatos, R. and Wangsness, P.J. 1981. Diurnal variations in plasma insulin and growth hormone associated with two stages of lactation in high producing dairy cows. Endocrinol. 108, 300-304.

Verde, L.S. and Trenkle, A. 1987. Concentrations of hormones in plasma from cattle with different growth potentials. J. Anim. Sci. 64, 426-432.

Wheaton, J.E., Al-Raheem, S.N., Massri, Y.G. and Marcek, J.M. 1986. Twenty-four-hour growth hormone profiles in angus steers. J. Anim. Sci. 62 1267-1272.

INFLUENCE OF SOMATOTROPIN ON METABOLISM

R.G. Vernon

Hannah Research Institute
Department of Biochemistry and Molecular Biology
Ayr, Scotland

ABSTRACT

The effects of somatotropin (ST) on metabolism in the whole animal and in specific organs are described. ST has two basic effects on metabolism: stimulation of productive processes (growth, milk synthesis) and promotion of homeorhetic adaptations which help provide the nutrients required for the productive processes. Effects of ST on glucose and fatty acid metabolism and their inter-relations have been defined in greatest detail. In many respects adaptations in response to treatment with ST resemble adaptations occurring during early lactation when endogenous plasma ST concentration is increased naturally.

INTRODUCTION

Somatotropin (ST) has many metabolic effects, in fact the hormone appears to be able to modify almost all aspects of metabolism (Table 1)! A pattern begins to emerge, however, when one considers that somatotropin has two basic effects: one concerned with productive processes which is probably partly but not exclusively mediated by IGF-I, and a second effect on nutrient supply which is probably mediated by somatotropin itself (Fig. 1). The nature and importance of these effects depends on the physiological state of the animal. For example in growing animals treatment with ST can lead to increased accretion of muscle protein (Boyd and Bauman, 1988) whereas during lactation it can lead to a loss of body protein (Peel et al., 1981). Effects of ST treatment depend also on the nutritional status of the animal. For example if an animal remains in positive energy balance during treatment with ST there will be diminished lipogenesis in adipose tissue but no increase in lipolysis, but if the animal moves into negative energy balance as a result of ST treatment then lipolysis will be enhanced also (Boyd and Bauman, 1988). It is a feature of ST action that it can move an animal into a transient phase of negative energy balance, promoting a productive effect without any increase in food intake initially, achieving this by causing a re-arrangement of metabolism elsewhere in the body to provide the necessary nutrients for the productive effect. Such a phenomenon occurs naturally during early lactation.

The importance of the homeorhetic action of ST is emphasised by the finding that it does not have a direct effect on appetite or digestion of carbohydrates, proteins and lipids in either growing (Boyd and Bauman, 1988) or lactating animals (Peel and Bauman, 1987; Chilliard, 1988; Hart, 1988). Prolonged treatment with ST ultimately leads to an increase in food intake but this is clearly a secondary response. However, ST treatment does increase calcium absorption (Boyd and Bauman, 1988). This is not surprising for whereas it is possible to increase the availability of organic nutrients by homeorhetic manipulation of metabolism, it is not possible to increase calcium availability for net bone growth other than by increasing absorption of dietary calcium.

TABLE 1 Some metabolic effects of somatotropin

Carbohydrate metabolism	Glucose production	↑
	Glucose oxidation	↓
Lipid metabolism	Fatty acid synthesis	↓
	Fatty acid oxidation	↑
Protein metabolism	Nitrogen retention	↑
	Protein synthesis	↑
Mineral metabolism	Ca^{2+} absorption	↑
	Ca^{2+} accretion	↑
Nucleic acid metabolism	Cell proliferation	↑

↑ ↓ Rate increased or decreased respectively

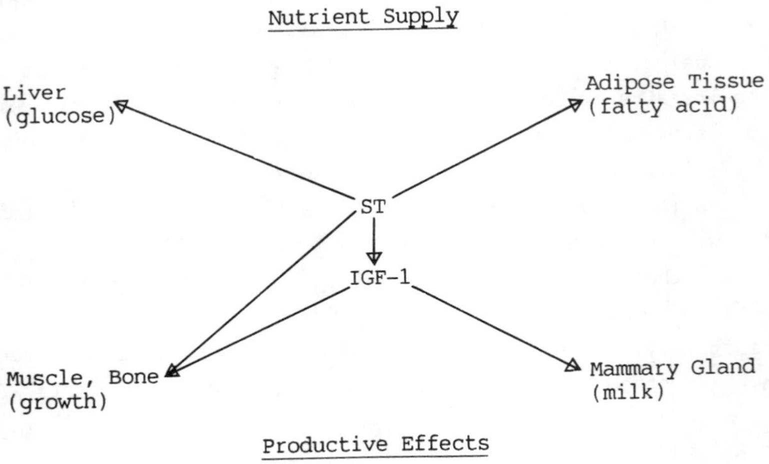

Nutrient Supply

Liver (glucose) → ST → Adipose Tissue (fatty acid)
ST → IGF-1
Muscle, Bone (growth) ← → Mammary Gland (milk)

Productive Effects

Fig. 1 Co-ordination of metabolism by somatotropin (ST)

Thus with the exception of calcium, effects of ST on nutrient supply in the short-term are through the homeorhetic manipulation of metabolism. In subsequent sections I will discuss the effects of ST on metabolism in the whole animal, then in liver, adipose tissue and muscle in relation to nutrient supply and finally in muscle, bone and the mammary gland in relation to nutrient utilisation.

METABOLISM IN THE WHOLE ANIMAL

Increased milk production in response to ST treatment requires additional glucose for lactose synthesis in the mammary gland, and indeed several studies have revealed an increase in glucose irreversible loss in ST-treated animals during lactation and also in growing pigs (Table 2). Curiously no change in glucose irreversible loss was found in one study with lactating cows and another with lactating sheep (Table 2); the reason for this is not certain, but it is possible that increased glucose uptake by the mammary gland for lactose synthesis was balanced by decreased glucose oxidation to CO_2 in the animal as a whole. Bauman et al. (1988) have shown that ST-treatment can lead to decreased glucose oxidation to CO_2 along with a concomitant increase in oxidation of FFA (unesterified fatty acids) (Table 2).

Increased FFA irreversible loss in response to ST treatment has been observed in several studies, mostly with lactating animals but also in growing heifers (Table 2). No such increase was found in one study with cows at peak lactation but the FFA irreversible loss was already very high in these animals prior to ST treatment (McDowell et al., 1987a). Taken together these various studies suggest that ST treatment can make more glucose available for use by the mammary gland by increasing glucose production and by decreasing glucose utilisation for oxidative and perhaps other purposes in the animal, the latter being achieved at least in part by increased oxidation of FFA. It has long been known from studies with laboratory species that increased availability of fatty acid will in itself lead to decreased glucose oxidation in peripheral tissues through inhibition of pyruvate dehydrogenase activity caused at least in part by the products of fatty acid oxidation (Denton et al., 1978). The effects of ST treatment on pyruvate dehydrogenase activity have yet to be investigated, but lactation in sheep, which results in increased endogenous ST (Vernon et al., 1981), leads to decreased pyruvate dehydrogenase activity in muscle and adipose tissue (Vernon et al., 1987). It is pertinent to note that in all of the studies summarised in Table 2 with the exception of the one study with growing pigs, ST treatment was for only 5 to 14 days, that is measurements would be made during a period when production would be increased without any increase in food intake. Bauman and colleagues (Tyrrell et al., 1988) indeed showed that in their studies with lactating cows the animals went into negative energy balance as a result of ST treatment. It is not known what happens to glucose and fatty acid metabolism in animals after prolonged treatment with ST (i.e. when food intake has increased); as plasma FFA concentrations are not normally elevated in such animals (Boyd and Bauman, 1988) it would seem unlikely that increased fatty acid oxidation would be occurring. The above discussion and results of Table 2 thus probably apply to that key transient initial period when ST promotes increased production via homeorhetic mechanisms.

TABLE 2   Metabolic effects of somatotropin (ST) in the whole animal

State / Animal / ST treatment	Peak lactation cow 6 d[a]	Mid lactation cow 6 d[a]	Mid lactation cow 14 d[b]	Lactating sheep 5 d[c]	Growing heifer 14 d[d]	Growing pig 28 d[e]
Glucose IRL	↑	—	↑	—		↑
Glucose → CO_2			↓			
% CO_2 from glucose			↓	—		
Fatty acid IRL	—	↑	↑	↑	↑	
Fatty acid → CO_2			↑		↑	
% CO_2 from fatty acid			↑	↑	↑	
Urea IRL	—	—				
Acetate IRL	—	—				
Leucine IRL					—	
Leucine → CO_2					↓	
Leucine → protein					↑	

IRL = Irreversible loss rate ↑ , ↓ , — , Increase, decrease or no change in response to ST. Results from (a) McDowell et al., 1987a; (b) Bauman et al., 1988; (c) McDowell et al., 1985; (d) Eisemann et al., 1986a; (e) Gopinath and Etherton, 1988.

In contrast to glucose and FFA, treatment of lactating cows with ST had no effect on either urea or acetate irreversible loss, at least over a 6-day treatment period (Table 2). A lack of effect on urea turnover is surprising as treatment with ST usually increases nitrogen retention (Hart and Johnson, 1986). ST-treatment had no effect on leucine irreversible loss in growing heifers, but further analysis showed that leucine oxidation to CO_2 was decreased whereas protein synthesis from leucine increased (Table 2).

ENDOCRINE INTERACTIONS

Administration of ST results in a marked increase in serum IGF-I (insulin-like growth factor-I) (see preceding Chapters) but usually has no effect on serum insulin concentration in lactating animals (Peel et al., 1982, 1983; McDowell et al., 1987b; Davis et al., 1987, 1988a; Soderholm et al., 1988) although an increase has been found in some studies (Hart et al., 1985; McDowell et al., 1987a). In contrast studies with growing pigs (Wray-Cahen et al., 1987; Gopinath and Etherton, 1988; Novakofski et al., 1988), sheep (Johnsson and Hart, 1985; Johnsson et al., 1987) and cattle (Peters, 1986; Eisemann et al., 1986a) have found an increase in serum insulin, although no effect was seen in one study with growing cattle (Crooker et al., 1988). Treatment with ST has no apparent effect on the serum concentration of a number of hormones including prolactin, glucagon, cortisol, thyroxine and triiodothyronine, at least in lactating cows (Peel et al., 1982, 1983).

Despite its effects on insulin concentration, treatment with ST decreases the ability of insulin to increase glucose utilisation; this has been shown by glucose tolerance tests in growing pigs (Gopinath and Etherton, 1988; Novakofski et al., 1988) and sheep (Hart et al., 1984) and in lactating sheep (Leenanuruska and McDowell, 1987), and also by the more rigorous glucose clamp procedure in growing pigs (Wray-Cahen et al., 1987). This impairment of insulin action should also facilitate the preferential utilisation of glucose by the mammary gland, for in lactating ruminants the mammary gland is insensitive to insulin.

In contrast to its effects on insulin action on glucose metabolism, ST treatment has been reported to increase the antilipolytic effect of insulin in vivo in pigs (Wray-Cahen, 1987; Novakofski et al., 1988) and lactating cows (Sechen et al., 1988)! The physiological advantage of this is not obvious.

HOMEORHETIC EFFECTS - NUTRIENT SUPPLY

Liver metabolism

In ruminants increased glucose irreversible loss implies increased glucose production, primarily by the liver. In the short-term (i.e. over a matter of hours) this could be the result of increased glycogenolysis, but over a number of days it must be the result of enhanced gluconeogenesis. Increased hepatic glucose output in vivo in response to ST treatment has not been demonstrated unequivocally but such a trend has been reported in lactating cows (Cohick et al., 1987) and in growing sheep (Elcock et al., 1988). Direct evidence for increased gluconeogenesis has been obtained from in vitro studies with liver from ST-treated lactating cows, while hepatic glycogen concentration was not altered by ST treatment in these animals (Pocius and Herbein, 1986). The liver has receptors for ST (Chilliard, 1988) hence these effects on glucose metabolism are probably direct (i.e. not mediated by IGF-I).

In contrast to the consistent and predictable changes, or at least trends, in hepatic carbohydrate metabolism in response to ST treatment, changes in hepatic lipid metabolism are neither clear not consistent. The uptake of fatty acids by the liver is thought to vary with plasma FFA concentration (Bell, 1980), hence in situations where ST treatment leads to increased plasma FFA concentration increased hepatic uptake is to be expected; such a trend has been found in studies with growing sheep (Leenanuruska et al., 1986; Elcock et al., 1988). The fate of this fatty acid, however, is far from clear. Within the hepatocyte, fatty acids may be esterified or transported into the mitochondria via the action of carnitine palmitoyl CoA acyltransferase (Zammit, 1984) (Fig. 2).

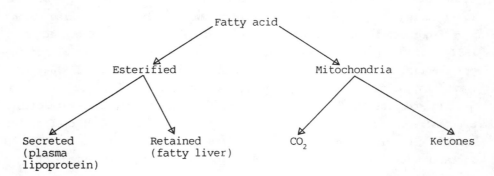

Fig. 2   Liver fatty acid metabolism

Esterified fatty acid is normally secreted as lipoprotein lipid (mostly triacylglycerol) or can be retained in the liver (leading to the pathological condition of 'fatty liver'). Within the mitochondria fatty acid may be oxidised to CO_2 or converted to ketones (acetoacetate and β-hydroxybutyrate) which are released into the blood for use by peripheral tissues including the mammary gland (Bell, 1980; Zammit, 1984).

No evidence of lipid accumulation was found in livers of ST treated lactating cows (Pocius and Herbein, 1986) while no change in hepatic lipoprotein lipid secretion was found in vivo in growing calves (Niumsup et al., 1985), whereas hepatic lipid secretion is claimed to decrease in lactating sheep (Niumsup et al., 1985)! There is a suggestion of increased oxidation of fatty acid to CO_2 from studies in vitro with liver slices from ST-treated lactating cows (this is to be expected if gluconeogenesis is increased) but no evidence for increased ketone production was found (Pocius and Herbein, 1986). A lack of an effect of ST treatment on ketogenesis is also suggested by absence of any change in plasma concentration of ketones in a number of studies with lactating cows (Pocius and Herbein, 1986; McDowell et al., 1987a, 1987b) growing heifers (Eisemann et al.,1986a; Leenanuruska et al., 1985) and lambs (Elcock et al., 1988; Lakehal et al., 1989). However, a marginal increase in hepatic ketone body output in vivo was noted in growing sheep treated with ST (Elcock et al., 1988). Thus although treatment of ST is likely to increase hepatic uptake of fatty acids, especially if it results in a period of negative energy balance, the fate of these fatty acids remains to be clarified.

Adipose tissue metabolism

Long-term treatment of animals with ST leads to a decrease in adiposity (Boyd and Bauman, 1988; Vernon and Flint, 1989). This could be the result of decreased lipid synthesis, increased lipolysis or a combination of both. Fatty acids required for lipid synthesis may be produced by lipogenesis (fatty acid synthesis) or released from plasma lipoproteins by the action of lipoprotein lipase. Some fatty acids released by lipolysis may also be re-esterified. Long-term treatment of pigs (Magri et al., 1987) sheep (Sinnett-Smith and Woolliams, 1989) or cattle (Skarda et al., 1989) with ST decreases the rate of lipogenesis in adipose tissue. Such an effect was not found in one study with steers (Peters, 1986), while a recent study shows a marked breed difference in the ability of ST treatment to decrease lipogenesis in adipose tissue from growing sheep (Sinnett-Smith and Wooliams, 1989). In addition maintenance of adipose tissue from sheep (Vernon, 1982; Vernon and Finley, 1988) pigs (Walton and Etherton, 1986; Walton et al., 1986) and cattle (Etherton et al., 1987) in tissue culture for 24-48h in the presence of ST also results in a decreased rate of lipogenesis, indicating that ST is acting on the tissue itself. ST receptors have been demonstrated in adipocytes from rats and man but their presence in adipocytes from farm animals has yet to be confirmed, whereas adipocytes lack IGF-I receptors (see Vernon and Flint, 1989). Studies with adipose tissue in vitro suggest that ST is acting primarily as an insulin antagonist (Vernon, 1982; Walton et al., 1986; Walton and Etherton, 1986; Etherton et al., 1987; Vernon and Finley, 1988) and have shown that one effect of ST is to antagonise the ability of insulin to activate acetyl CoA carboxylase (Vernon and Flint, 1989), the most important regulatory enzyme of lipogenesis. During early lactation the rate of lipogenesis in adipose tissue decreases to very low levels in domestic ruminants, recovering during later stages when animals return to positive energy balance (Vernon, 1988). Recovery of lipogenesis in adipose tissue from lactating sheep can be achieved in vitro by prolonged incubation with insulin and glucocorticoid; recovery is prevented by ST (Vernon and Finley, 1988).

Treatment of sheep with ST also decreased the rate of esterification but had no effect on lipoprotein lipase activity (Sinnett-Smith and Wooliams, 1989). However, during early lactation when endogenous ST is high in plasma, lipoprotein lipase activity is low in adipose tissue (Vernon, 1988). Also adipose tissue at this time is clearly insulin-resistant (Vernon and Taylor, 1988); it is not known if this is due to the elevated plasma ST concentration but this seems likely.

Treatment with ST can lead to increased plasma FFA concentrations, indicating enhanced lipolysis, although this is not always observed (Boyd and Bauman, 1988; Vernon and Flint, 1989). The mechanisms whereby ST increases lipolysis have not been fully resolved, studies being hampered by artefacts arising from impure preparations of ST (Boyd and Bauman, 1988). Some studies with recombinant or purified pituitary ST suggest that ST can increase lipolysis directly after a lag-phase of an hour or so both in vivo and in vitro, but most recent studies with such preparations have failed to reveal a direct lipolytic effect of ST in a variety of species either in vivo or in vitro (including sheep, pigs, cattle and even reindeer) (see Vernon and Flint, 1989). A very recent study with sheep adipose tissue suggests that lipolytic effects of ST may be mediated by IGF-I and, or, IGF-II (Lewis et al., 1988) but this awaits confirmation. Alternatively, or perhaps in addition, ST treatment can increase lipolysis indirectly by altering the responsiveness of adipocytes to other hormones. Treatment of steers (Peters, 1986) lactating cows (Sechen et al., 1985; McCutcheon and Bauman, 1986; Sechen et al., 1988) and growing pigs (Wray-Cahen et al., 1987; Novakofski et al., 1988; Brenner et al., 1989) with ST results in an increased surge in plasma glycerol and, or, fatty acid concentration in vivo in response to a catecholamine load; this effect was observed in growing steers (Peters, 1986) or lactating cows (Sechen et al., 1988) when animals were in either positive or negative energy balance. A detailed study by Bauman and colleagues (Sechen et al., 1988) showed that ST effected maximum response to catecholamine in vivo but not sensitivity (the concentration of catecholamine required to achieve half-maximum effect). In contrast, treatment of growing steers (Peters, 1986) or lambs (Sinnett-Smith and Wooliams, 1989) with ST had no effect on catecholamine stimulation of lipolysis in vitro. Evidence that ST can effect responsiveness of adipocytes to catecholamines also comes from studies with laboratory species (Vernon et al., 1987), whilst during early lactation the lipolytic effect of catecholamines is enhanced in cattle (Jaster and Wegner, 1981) and sheep (Vernon and Finley, 1985). Lactation also results in an increase in the number of β-adrenergic receptors of adipocytes in cattle (Jaster and Wegner, 1981) and sheep (Watt et al., 1989); this has been mimicked in sheep adipose tissue by incubating the tissue for 2 days with ST (Watt et al., 1989).

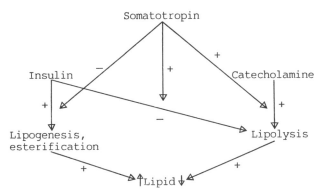

Fig. 3   Modulation of insulin and catecholamine action on adipose tissue by somatotropin. (+) stimulation; (−) inhibition; ↑ ↓ amount of lipid increased or decreased.

As well as these effects on the lipolytic actions of catecholamines, as noted already ST treatment also leads to an enhancement of the antilipolytic effect of insulin; ST may thus increase the responsiveness of both the lipolytic and antilipolytic signal transduction systems of adipocytes, rendering the whole system more responsive to changes in the hormonal milieu around the cell (Fig. 3).

Skeletal muscle: energy metabolism

There have been no studies on the effect of ST treatment on the energy metabolism of skeletal muscle itself, but several studies have explored the effects of ST treatment on the metabolism of the hind-limb in cattle and sheep; metabolism of the hind limb will be dominated by that of skeletal muscle although there will be some contribution from other tissues, especially adipose tissue. Treatment with ST results in increased blood flow to the hind-limb in lactating cows (McDowell et al., 1987b) and sheep (McDowell et al., 1985) and also growing heifers (Leenanuruska et al., 1985), thus increasing the supply of nutrients. Increased nutrient supply to muscle is obviously advantageous in growing animals but its purpose is less obvious in lactating animals. The arterio-venous difference of several nutrients across the hind-limb has been reported (Table 3) but with the exception of one study with growing lambs these studies have not made clear if there is a change in uptake when the increased blood flow is taken into consideration. It would appear however that in both lactating cows and in growing heifers and lambs glucose uptake by the hind-limb falls and, in cattle at least, lactate output increases (Table 3). This is consistent with a switch from glucose oxidation to fatty acid oxidation with a concomitant inhibition of pyruvate dehydrogenase as noted previously. The lactate will most probably recycle to the liver and be used for gluconeogenesis. The system is thus a device for conserving glucose.

There is no unequivocal evidence as yet for increased uptake of fatty acid by the hind-limb of lactating cattle or sheep in response to ST treatment; however, although non-significant, there was a tendency for the arterio-venous difference to increase in cattle which combined with an increased blood flow suggests increased uptake, while the study with growing lambs suggests a trend towards increased fatty acid uptake, although again this was not statistically significant (Lakehal et al., 1989). The picture is potentially complicated of course by the possibility of fatty acid uptake by muscle being partly balanced by increased fatty acid release by adipose tissue depots in the hind-limb in response to ST treatment (this would account for the apparent switch from fatty acid uptake to output seen in growing heifers (Leenanuruska et al., 1985).

TABLE 3 Effects of somatotropin treatment on the utilization of nutrients by the hind-limb in vivo

	Lactating		Growing	
	Cows a-v	Sheep a-v	Heifers a-v	Lambs uptake
Glucose	↓	—	↓	↓
Fatty acids	—	—		—
Acetate	—	—		—
β-hydroxybutyrate	—	—	—	—
Lactate (output)	↑	↑	↑	—

a-v, arterio-venous difference; ↑ , ↓ , — , decrease, increase or no change in response to somatotropin treatment. Results from:
a. McDowell et al., 1987b; b. McDowell et al., 1985; c. Leenanuruska et al., 1985; d. Lakehal et al., 1989.

Acetate and β-hydroxybutyrate are also oxidised by muscle but there was no change in uptake in growing sheep (Lakehal et al., 1989) while in lactating cows, small, non-significant falls in arterio-difference would be compensated by the increased blood flow hence a change in uptake seems unlikely.

PRODUCTIVE EFFECTS IN GROWING ANIMALS

Skeletal muscle: protein metabolism

Long-term treatment with ST increases muscle growth; there is increased cellular proliferation (probably mediated by IGF-I) and increased protein accretion (see Boyd & Bauman, 1988). In accordance with this ST treatment leads to increased nitrogen retention (Hart and Johnsson, 1986) and protein synthesis in the whole animal (Eisemann et al., 1986a), although some recent studies failed to find any effect on urea irreversible loss (Table 2). Further studies with both growing steers (Eisemann et al., 1986b) and lambs (Pell and Bates, 1987) have confirmed that ST treatment increases muscle protein synthesis but has no effect on protein degradation (Pell and Bates, 1987). The RNA concentration of skeletal muscle was increased by ST treatment in growing lambs (this is thought to reflect the number of ribosomes and hence the capacity for protein synthesis) and in addition there was a marginal increase in protein synthesis per g RNA, suggesting an increase in the efficiency of protein synthesis (Pell and Bates, 1987).

Buttery and colleagues have explored the effects of ST and IGF-I on protein metabolism in muscle cell cultures. With cells from foetal lambs (Harper et al., 1987) and adult sheep (Roe et al., 1989) they found that ST itself had no effect on either protein synthesis or degradation, whereas IGF-I stimulated protein synthesis and had a marginal inhibitory effect on degradation. These studies strongly suggest that the effects on ST on protein metabolism in muscle may be mediated by IGF-I.

It is pertinent to remember that whereas ST treatment promotes protein accretion in growing animals, it can lead to a loss of muscle protein in lactating animals, that is when dietary amino acid availability is inadequate to meet the increased milk protein production (see Introduction).

Bone: mineral metabolism

ST treatment also stimulates bone growth, promoting prechondrocyte differentiation and chrondrocyte proliferation; the latter effect is probably mediated by IGF-I (see Boyd and Bauman, 1988). In a detailed study with sheep, Braithwaite (1975) found that ST treatment stimulated intestinal calcium absorption, and the rates of both accretion and resorption of bone calcium, but with the net effect of increased deposition of calcium. The mechanism involved is uncertain but studies with hypophysectomised rats and also ST-deficient children show that ST treatment can increase plasma 1,25 dihydroxy vitamin D_3 concentration (the active form of the vitamin) which has a key role in control of calcium metabolism (see Boyd and Bauman, 1988).

PRODUCTIVE EFFECTS IN LACTATING ANIMALS

Mammary gland metabolism

Stimulation of milk production by ST has obvious consequences for mammary metabolism - synthetic processes must be increased in the gland and hence nutrient uptake must also be increased. ST treatment increases mammary blood flow in cows (Fleet et al., 1986; McDowell et al., 1987b; Davis et al., 1988a), sheep, (Fleet et al., 1988) and goats (Mepham et al., 1984) and hence nutrient supply: however, it is pertinent to note, that the effects of ST on milk yield are not solely due to increased availability of nutrients (Peel and Bauman, 1987), although this undoubtedly is a key part of the mechanism. ST treatment had no effect on the arterio-venous difference for glucose, acetate, oxygen or (essential) amino acids, but did increase uptake of all of these (blood flow is increased) (Table 4).

TABLE 4 Nutrient utilization by the mammary gland during lactation: response to somatotropin treatment

	Cows		Goats		Sheep	
	a-v	uptake	a-v	uptake	a-v	uptake
Glucose	—	↑	—	↑	—	↑
Amino acids	—	↑	—			
Acetate	—	↑			—	
β-hydroxybutyrate	—	↑				
Oxygen	—	↑			↓	
CO_2 (output)	↓				—	

a-v, ↑ , ↓ , — - see legend to Table 3.
Data from: cows, Fleet et al., 1986; McDowell et al., 1987b; Davis et al., 1988b and Fullerton, et al., 1989; goats, Mepham et al., 1984; sheep, Fleet et al., 1988.

The picture with regards to fatty acid metabolism is less clear: FFA may be taken up and also the fatty acids of blood triacylglycerols following hydrolysis by the action of lipoprotein lipase; the latter is thought to be the major source of exogenous fatty acid (Moore and Christie, 1979). One study with lactating cows revealed increased uptake of triacylglycerol fatty acids but no change in uptake of FFA (Davis et al., 1988b); whereas others have reported increased uptake of FFA in cows (McDowell et al., 1987b) and a decrease in the arterio-venous difference, at least, for triacylglycerol fatty acid in sheep (Niumsup et al., 1985)! Increased uptake of fatty acids from either source has potential rammifications for lipogenesis and can lead to a decrease in the proportion of fatty acids synthesised de novo in milk fat (Bitman et al.,

1984; Eppard et al., 1985). In lactating sheep at least there is evidence for an increased proportion of mammary CO_2 being derived from fatty acid on ST treatment; the proportion of CO_2 derived from glucose did not change (McDowell et al., 1985), however, but a large proportion of this is likely to be the result of glucose metabolism via the pentose phosphate pathway rather than via the tricarboxyllic acid cycle (Chaiyabutr et al., 1982). In accordance with a switch to increased fatty acid oxidation, mammary RQ fell in lactating cows (Fleet et al., 1986) and sheep (Fleet et al., 1988) on ST treatment and in cattle there is a suggestion of a switch from lactate uptake to output by the mammary gland with ST treatment (McDowell et al., 1987b).

The effects of ST on mammary metabolism are thought to be indirect as no receptors for ST appear to be present on mammary epithelial cells in cattle (Akers, 1985; Peel and Bauman, 1987; Chilliard, 1988) and incubation in tissue culture of mammary tissue in the presence of ST has no effect on metabolism (Skarda et al., 1982; Goodman et al., 1983; Gertler et al., 1983; Baumrucker, 1985). Furthermore, whereas IGF-I is mitogenic for bovine mammary tissue, it has no metabolic effects other than those associated with proliferation (Shamay et al., 1988). The mechanisms whereby ST increases milk yield (in addition to providing nutrients) are thus unresolved. Nevertheless as far as mammary metabolism is concerned, the picture appears uncomplicated, with increased uptake of precursors such as glucose, amino acids and fatty acids and increased synthesis of milk constituents with some proportionate increase in exogenous fatty acid utilisation both for milk fat synthesis and for oxidation.

SUMMARY

There is still a paucity of information on the metabolic effect of ST on metabolism in farm animals; much of the data noted in this review comes from preliminary communications and some is derived from adaptations occurring in animals during early lactation when endogenous ST in plasma is elevated (the similarities between the adaptations of ST-treated animals and of high-yielding cows have been emphasised previously, Peel and Bauman, 1987; Hart, 1988). It is also clear that some studies have produced conflicting conclusions. Nevertheless a reasonable picture of metabolic adaptations is beginning to appear, especially with regards to glucose and fatty acid metabolism and their interactions. It must be emphasised that most of the studies described above have been relatively short-term and deal with the intriguing situation when ST treatment is promoting 'production' without concomitant increase in food intake. During this period animals are likely to move into negative energy balance and so will be mobilising lipid. This fatty acid not only supplies an energy source and in the lactating animal a source of precursor for milk fat, but also provides a means of diminishing glucose oxidation by muscle and other tissues, facilitating the livers efforts to provide additional glucose for milk synthesis. For the lactating animal the advantage of conserving glucose is obvious but for the growing animal it is less clear.

Prolonged treatment with ST results in increased food intake hence increased milk yield or growth can be fuelled by extra dietary nutrients and so the need to mobilise fat reserves is lost. Some of the metabolic adaptations (increased fatty acid oxidation?) observed during shorter periods of treatment described above thus may be not be observed. Decreased lipogenesis in adipose tissue is still probable and insulin-resistance with respect to glucose metabolism may still serve to decrease glucose oxidation by peripheral tissues, hence even when nutrient availability is increased as a result of greater intake, ST may still have a useful role in manipulating metabolism to ensure that nutrients are directed to specific organs and that unnecessary glucose utilisation (e.g. for oxidation) is restricted.

REFERENCES

Akers, R.M. 1985. Lactogenic hormones: binding sites, mammary growth, secretory cell differentiation, and milk biosynthesis in ruminants. J. Dairy Sci., 68, 501-519.
Bauman, D.E., Peel, C.J., Steinhour, W.D., Reynolds, P.J., Tyrrell, H.F., Brown, A.C.G. and Haaland, G.L. 1988. Effect of bovine somatotropin on metabolism of lactating dairy cows: influence on rates of irreversible loss and oxidation of glucose and nonesterified fatty acids. J. Nutr., 118, 1031-1040.
Baumrucker, C.R. 1985. Growth hormone does not directly affect bovine mammary tissue growth nor lactating acini milk production in culture. J. Dairy Sci., 69, suppl. 1. p.106.
Bell, A.W. 1980. Lipid metabolism in liver and selected tissues and in the whole body of ruminant animals. Prog. Lipid Res., 18, 117-164.
Bitman, J., Wood, D.L., Tyrrell, H.F., Bauman, D.E., Peel, C.J., Brown, A.C.G. and Reynolds, P.J. 1984. Blood and milk lipid responses induced by growth hormone administration in lactating cows. J. Dairy Sci., 67, 2873-2880.
Boyd, R.D. and Bauman, D.E. 1988. Mechanism of action of somatotropin in growth. In "Current Concepts of Animal Growth" (Ed. D.R. Campion, G.J. Hausman and R.J. Martin). (New York), in press.
Braithwaite, G.D. 1975. The effect of growth hormone on calcium metabolism in the sheep. Br. J. Nutr., 33, 309-314.

Brenner, K.-V., Novakofski, J., Bechtel, P.J. and Easter, R.A. 1989. Metabolic and endocrine challenge of somatotropin treated pigs. In "Biotechnology in Growth Regulation" (Ed. R.B. Heap, C.G. Prosser and G.E. Lamming). (Butterworths, London). (in press).

Chilliard, Y. 1988. Roles et mecanismes d'action de la somatotropine (hormone de croissance) chez le ruminant en lactation. Reprod. Nutr. Develop., 28, 39-59.

Chaiyabutr, N., Faulkner, A. and Peaker, M. 1983. Effects of exogenous glucose on glucose metabolism in the lactating goat in vivo. Br. J. Nutr., 49, 159-165.

Cohick, W.S., Slepetis, R., Plaut, K. and Bauman, D.E. 1987. Effect of exogenous somatotropin on serum somatomedin-C (SmC) and hepatic metabolism of lactating cows. J. Anim. Sci., 65, suppl. 1, p.248.

Crooker, B.A., Bauman, D.E., Cohick, W.S. and Harkins, W. 1988. Effect of dose of exogenous bovine somatotropin on nutrient utilization by growing dairy heifers. J. Anim. Sci., 66, suppl. 1, p.299.

Davis, S.R., Collier, R.J., McNamara, J.P., Head, H.H. and Sussman, W. 1988a. Effects of thyroxine and growth hormone treatment of dairy cows on milk yield, cardiac output and mammary blood flow. J. Anim. Sci., 66, 70-79.

Davis, S.R., Collier, R.J., McNamara, J.P., Head, H.H., Croom, W.J. and Wilcox, C.J. 1988b. Effects of thyroxine and growth hormone treatment of dairy cows on mammary uptake of glucose, oxygen and other milk fat precursors. J. Anim. Sci., 66, 80-89.

Davis, S.R., Gluckman, P.D., Hart, I.C. and Henderson, H.V. 1987. Effects of injecting growth hormone or thyroxine on milk production and blood plasma concentrations of insulin-like growth factors I and II in dairy cows. J. Endocr., 114, 17-24.

Denton, R.M., Hughes, W.A., Bridges, B.J., Brownsey, R.W., McCormack, J.G. and Stansbie, D. 1978. Regulation of mammalian pyruvate dehydrogenase by hormones. Horm. Cell. Regul., 2, 191-208.

Eisemann, H.J., Hammond, A.C., Bauman, D.E., Reynolds, P.J., McCutcheon, S.N., Tyrrell, H.F. and Haaland, G.L. 1986a. Effect of bovine growth hormone administration on metabolism of growing hereford heifers: protein and lipid metabolism and plasma concentrations of metabolites and hormones. J. Nutr., 116, 2504-2515.

Eisemann, J.H., Hammond, A.C., Rumsey, T.S. and Bauman, D.E. 1986b. Tissue protein synthesis rates in beef steers injected with placebo or bovine growth hormone. J. Anim. Sci., 63, suppl. 1, p.217.

Elcock, C., Buttle, H.L., Coles, H.J., Hathorn, D.J., Simmonds, A.D. and Pell, J.M. 1988. The effect of growth hormone on nutrient partitioning in lambs. J. Endocr., 117, suppl. 1, p.59.

Eppard, P.J., Bauman, D.E., Bitman, J., Wood, D.L., Akers, R.M. and House, W.A. 1985. Effect of dose of bovine growth hormone on milk composition: α-lactalbumin, fatty acids, and mineral elements. J. Dairy Sci., 68, 3047-3054.

Etherton, T.D., Evock, C.M. and Kensinger, R.S. 1987. Native and recombinant bovine growth hormone antagonize insulin action in cultured bovine adipose tissue. Endocrinology 121, 699-703.

Fleet, I.R., Fullerton, F.M., Heap, R.B., Mepham, T.B., Gluckman, P.D. and Hart, I.C. 1988. Cardiovascular and metabolic responses during growth hormone treatment of lactating sheep. J. Dairy Res., 55, 479-485.

Fleet, I.R., Fullerton, F.M. and Mepham, T.B. 1986. Effects of exogenous growth hormone on mammary function in lactating Jersey cows. J. Physiol., 376, 19P.

Fullerton, F.M., Mepham, T.B., Fleet, I.R. and Heap, R.B. 1989. Changes in mammary uptake of essential amino acids in lactating jersey cows in response to exogenous bovine pituitary somatotropin. In

"Biotechnology in Growth Regulation" (Ed. R.B. Heap, C.G. Prosser and G.E. Lamming). (Butterworths, London). (in press).

Gertler, A., Cohen, N. and Maoz, A. 1983. Human growth hormone but not ovine or bovine growth hormones exhibits galactopoietic prolactin-like activity in organ culture from bovine lactating mammary gland. Mol. Cell. Endocr., 33, 169-182.

Goodman, G.T., Akers, R.M., Friderici, K.H. and Tucker, H.A. 1983. Hormonal regulation of α-lactalbumin secretion from bovine mammary tissue cultured in vitro. Endocrinology 112, 1324-1330.

Gopinath, R. and Etherton, T.D. 1988. Porcine growth hormone (pGH) increases hepatic glucose production rate and impairs glucose clearance. J. Anim. Sci., 66, suppl. 1, p.292.

Hart, I.C. 1988. Altering the efficiency of milk production of dairy cows with somatotropin. In "Nutrition and Lactation in the Dairy Cow" (Ed. P.C. Garnsworthy). (Butterworths, London), pp. 232-248.

Hart, I.C., Chadwick, P.M.E., Boone, T.C., Langley, K.E., Rudman, C. and Souza, L.M. 1984. A comparison of the growth-promoting, lipolytic, diabetogenic and immunological properties of pituitary and recombinant-DNA-derived bovine growth hormone (somatotropin). Biochem. J., 224, 93-100.

Hart, I.C., Chadwick, P.M.E., James, S. and Simmonds, A.D. 1985. Effect of intravenous bovine growth hormone or human pancreatic growth hormone-releasing factor on milk production and plasma hormones and metabolites in sheep. 1985. J. Endocr., 105, 189-196.

Hart, I.C., and Johnsson, I.D. 1986. Growth hormone and growth in meat producing animals. In "Control and Manipulation of Animal Growth (Eds. P.J. Buttery, D.B. Lindsay and N.B. Haynes). (Butterworths, London). pp. 135-159.

Harper, J.M.M., Soar, J.B. and Buttery, J.P. 1987. Changes in protein metabolism of ovine primary muscle cultures on treatment with growth hormone, insulin, insulin-like growth factor I or epidermal growth factor. J. Endocr., 112, 87-96.

Jaster, E.H. and Wegner, T.N. 1981. Beta-adrenergic receptor involvement in lipolysis of dairy cattle subcutaneous adipose tissue during dry and lactating state. J. Dairy Sci., 64, 1655-1663.

Johnsson, I.D. and Hart, I.C. 1985. The effects of exogenous bovine growth hormone and bromocriptine on growth, body development, fleece weight and plasma concentrations of growth hormone, insulin and prolactin in female lambs. Anim. Prod., 41, 207-217.

Johnsson, I.D., Hathorn, D.J., Wilde, R.M., Treacher, T.T. and Butler-Hogg, B.W. 1987. The effect of dose and method of administration of biosynthetic bovine somatotropin on live-weight gain, carcass composition and wool growth in young lambs. Anim. Prod., 44, 405-414.

Lakehal, F., Crompton, L.A. and Lomax, M.A. 1989. The effect of growth hormone on hind-limb muscle metabolism. In "Biotechnology in Growth Regulation" (Ed. R.B. Heap, C.G. Prosser and G.E. Lamming). (Butterworths, London). (in press).

Leenanuruksa, D. and McDowell, G.H. 1987. Diabetogenic effects of exogenous growth hormone demonstrated in vivo in sheep. Proc. Nutr. Soc. Aust., 12, p175.

Leenanuruksa, D., Niumsup, P., van der Walt, J.G., Gooden, J.M. and McDowell, G.H. 1986. Effect of exogenous growth hormone on hepatic-exchanges of glucose, free fatty acids and insulin in lactating ewes. Proc. Nutr. Soc. Aust., 11, p95.

Leenanuruksa, D., Smithard, R., McDowell, G.H., Gooden, J.M., Jois, M. and Niumsup, P. 1985. Effects of growth hormone on nutrient utilisation and blood flow in muscle tissue of growing calves. Proc. Nutr. Soc.

Aust., 10, p152.
Lewis, K.J., Molan, P.C., Bass, J.J. and Gluckman, P.D. 1988. The lipolytic activity of low concentrations of insulin-like growth factors in ovine adipose tissue. Endocrinology 122, 2554-2557.
Magri, K.A., Gopinath, R. and Etherton, T.D. 1987. Inhibition of lipogenic enzyme activities by porcine growth hormone (pGH). J. Dairy Sci., 65, Suppl. 1, p.258.
McCutcheon, S.N. and Bauman, D.E. 1986. Effect of chronic growth hormone treatment on responses to epinephrine and thyrotropin-releasing hormone in lactating cows. J. Dairy Sci., 69, 44-51.
McDowell, G.H., Hart, I.C., Bines, J.A., Lindsay, D.B. and Kirby, A.C. 1987a. Effects of pituitary-derived bovine growth hormone on production parameters and biokinetics of key metabolites in lactating dairy cows at peak and mid-lactation. Aust. J. Biol. Sci., 40, 191-202.
McDowell, G.H., Gooden, J.M., Leenanuruksa, D., Jois, M. and English, A.W. 1987b. Effects of exogenous growth hormone on milk production and nutrient uptake by muscle and mammary tissues of dairy cows in mid-lactation. Aust. J. Biol. Sci., 40, 295-306.
McDowell, G.H., Gooden, J.M., van der Walt, J.G., Smithard, R., Leenanuruksa, D. and Niumsup, P. 1985. Metabolism of glucose and free fatty acids in lactating ewes treated with growth hormone. Proc. Nutr. Soc. Aust., 10, p155.
Mepham, T.B., Lawrence, S.E., Peters, A.R. and Hart, I.C. 1984. Effects of exogenous growth hormone on mammary function in lactating goats. Horm. Metab. Res., 16, 248-253.
Moore, J.H. and Christie, W.W. 1979. Lipid metabolism in the mammary gland of ruminant animals. Prog. Lipid Res., 17, 347-395.
Niumsup, P., McDowell, G.H., Leenanuruksa, D. and Gooden, J.M. 1985. Plasma triglyceride metabolism in lactating ewes and growing calves treated with growth hormone. Proc. Nutr. Soc. Aust., 10, p154.
Novakofski, J., Brenner, K., Easter, R., McLaren, D. Jones, R., Ingle, D. and Bechtel, P. 1988. Effects of porcine somatotropin on swine metabolism. Fed. Proc. Amer. Soc. Exptl. Biol., 2, p848.
Peel, C.J. and Bauman, D.E. 1987. Somatotropin and lactation. J. Dairy Sci., 70, 474-486.
Peel, C.J., Bauman, D.E., Gorewit, R.C. and Sniffen, C.J. 1981. Effect of exogenous growth hormone on lactational performance in high yielding dairy cows. J. Nutr., 111, 1662-1671.
Peel, C.J., Fronk, T.J., Bauman, D.E. and Gorewit, R.C. 1982. Lactational response to exogenous growth hormone and abomasal infusion of a glucose-sodium caseinate mixture in high-yielding dairy cows. J. Nutr., 112, 1770-1778.
Peel, C.J., Fronk, T.J., Bauman, D.E. and Gorewit, R.C. 1983. Effect of exogenous growth hormone in early and late lactation on lactational performance of dairy cows. J. Dairy Sci., 66, 776-782.
Pell, J.M. and Bates, P.C. 1987. Collagen and non-collagen protein turnover in skeletal muscle of growth hormone-treated lambs. J. Endocr., 115, R1-R4.
Peters, J.P. 1986. Consequences of accelerated gain and growth hormone administration for lipid metabolism in growing beef steers. J. Nutr., 116, 2490-2503.
Pocius, P.A. and Herbein, J.H. 1986. Effects of in vivo administration of growth hormone on milk production and in vitro hepatic metabolism in dairy cattle. J. Dairy Sci., 69, 713-720.
Roe, J.A., Heywood, C.M., Harper, J.M.M. and Buttery, P.J. 1989. Comparative aspects of selected hormones and growth factors on protein metabolism in adult and foetal ovine primary muscle cultures.

In "Biotechnology in Growth Regulation" (Ed. R.B. Heap, C.G. Prosser and G.E. Lamming). (Butterworths, London). (In press).

Sechen, S.J., Dunshea, F.R. and Bauman, D.E. 1988. Mechanism of bovine somatotropin (bST) in lactating cows: effect on response to homeostatic signals (epinephrine and insulin). J. Dairy Sci., $\underline{71}$ Suppl. 1, p168.

Sechen, S.J., McCutcheon, S.N. and Bauman, D.E. 1985. Response to metabolic challenges in lactating dairy cows during short-term bovine growth hormone treatment. J. Dairy Sci., 68, Suppl. 1, p170.

Shamay, A., Cohen, N., Niwa, M. and Gertler, A. 1988. Effect of insulin-like growth factor I on deoxyribonucleic acid synthesis and galactopoiesis in bovine undifferentiated and lactating mammary tissue in vitro. Endocrinology $\underline{123}$, 804-809.

Sinnett-Smith, P.A. and Woolliams, J.A. 1989. Anti-lipogenic but not lipolytic effects of recombinant DNA-derived bovine somatotropin treatment on ovine adipose tissue; variation with genetic type. Int. J. Biochem. (in press).

Skarda, J., Krejci, P., Slaba, J. and Husakova, E. 1989. Blood constituents and subcutaneous adipose tissue metabolism of dairy cows after administration of recombinant bovine somatrotropin (bST) in a prolonged release formulation. In "Biotechnology in Growth Regulation" (Ed. R.B. Heap, C.G. Prosser and G.E. Lamming). (Butterworths, London). (In press).

Skarda, J., Urbanova, E., Becka, S., Houdebine, L.M., Delouis, C., Pichova, D., Picha, J. and Bilek, J. 1982. Effect of bovine growth hormone on development of goat mammary tissue in organ culture. Endocr. Expt., $\underline{16}$, 19-31.

Soderholm, C.G., Otterby, D.E., Linn, J.G., Ehle, F.R., Wheaton, J.E., Hansen, W.P. and Annexstad, R.J. 1988. Effects of recombinant bovine somatotropin on milk production, body composition and physiological parameters. J. Dairy Sci., $\underline{71}$, 355-365.

Tyrrell, H.F., Brown, A.C.G., Reynolds, P.J., Haaland, G.L., Bauman, D.E., Peel, C.J. and Steinhour, W.D. 1988. Effect of bovine somatotropin on metabolism of lactating dairy cows: energy and nitrogen utilization as determined by respiration calorimetry. J. Nutr., $\underline{118}$, 1024-1030.

Vernon, R.G. 1982. Effects of growth hormone on fatty acid synthesis in sheep adipose tissue. Int. J. Biochem., $\underline{14}$, 255-258.

Vernon, R.G. 1988. The partition of nutrients during the lactation cycle. In "Nutrition and Lactation in the Dairy Cow" (Ed. P.C. Garnsworthy). (Butterworths, London). pp. 32-52.

Vernon, R.G., Clegg, R.A. and Flint, D.J. 1981. Metabolism of sheep adipose tissue during pregnancy and lactation. Biochem. J., $\underline{200}$, 307-314.

Vernon, R.G., Faulkner, A., Finley, E., Pollock, H. and Taylor, E. 1987. Enzymes of glucose and fatty acid metabolism of liver, kidney, skeletal muscle, adipose tissue and mammary gland of lactation and non-lactating sheep. J. Anim. Sci., $\underline{64}$, 1395-1411.

Vernon, R.G. and Finley, E. 1985. Regulation of lipolysis during pregnancy and lactation in sheep. Biochem. J., $\underline{230}$, 651-656.

Vernon, R.G. and Finley, E. 1988. Roles of insulin and growth hormone in the adaptations of fatty acid synthesis in white adipose tissue during the lactation cycle in sheep. Biochem. J., $\underline{256}$, 873-878.

Vernon, R.G., Finley, E. and Flint, D.J. 1987. Role of growth hormone in the adaptations of lipolysis in rat adipocytes during recovery from lactation. Biochem. J., $\underline{242}$, 931-934.

Vernon, R.G. and Flint, D.J. 1989. Role of growth hormone in the regulation of adipocyte growth and function. In "Biotechnology in Growth Regulation (Ed. R.B. Heap, C.G. Prosser and G.E. Lamming).

(Butterworths, London). (in press).
Vernon, R.G. and Taylor, E. 1988. Insulin, dexamethasone and their interactions in the control of glucose metabolism in adipose tissue from lactating and non-lactating sheep. Biochem. J., 256, 509-514.
Walton, P.E. and Etherton, T.D. 1986. Stimulation of lipogenesis by insulin in swine adipose tissue: antagonism by porcine growth hormone. J. Anim. Sci., 62, 1584-1595.
Walton, P.E., Etherton, T.D. and Evock, C.M. 1986. Antagonism of insulin action in cultured pig adipose tissue by pituitary and recombinant porcine growth hormone: potentiation by hydrocortisone. Endocrinology 118, 2577-2581.
Watt, P.W., Clegg, R.A., Flint, D.J. and Vernon, R.G. 1989. Increases in sheep β-receptor number on exposure to growth hormone in vitro. In "Biotechnology in Growth Regulation" (Eds. R.B. Heap, C.G. Prosser and G.E. Lamming). (Butterworths, London). (in press).
Wray-Cahen, D., Boyd, R.D., Bauman, D.E., Ross, D.A. and Fagin, K. 1987. Metabolic effects of porcine somatotropin (pST) in growing swine. J. Anim. Sci., 65 Suppl. 1, p261.
Zammit, V.A. 1984. Mechanisms of regulation of the partition of fatty acids between oxidation and esterification in the liver. Prod. Lipid Res., 23, 39-67.

ALTERNATIVES TO GROWTH HORMONE FOR THE MANIPULATION OF ANIMAL PERFORMANCE

D.J. Flint

Hannah Research Institute,
Ayr, Scotland KA6 5HL

ABSTRACT

Recombinant growth hormones (GH) are currently able to stimulate milk production in dairy cows as well as improving growth performance in cattle sheep and pigs. A number of alternatives exist however, which may compete with and supersede GH. They include agents designed to manipulate endogenous GH such as growth hormone releasing factor (GRF), immunization against somatostatin, and the use of monoclonal antibodies to enhance GH bioactivity. The search for active "fragments" of GH has currently failed to produce peptides whose bioactivity is not considerably reduced compared with GH. Insulin like growth factor I (IGF-I), which is thought to be responsible for mediating certain effects of GH, as well as anti-idiotypic antibodies which mimic GH may permit the expression of specific actions of GH, rather than its full range of metabolic effects. Manipulation of carcass composition has also been achieved by the use of β agonists and by the use of antibodies to adipose tissue. The ultimate approach to manipulation of animal performance however must be the use of transgenic animals, where the GH gene has been microinjected into the fertilized egg and incorporated into the genome. The potential of this approach has been demonstrated in pigs although precise regulation of GH secretion remains a problem.

INTRODUCTION

Considerable interest is currently focussed on the possible commercial use of recombinant bovine and porcine growth hormone. Bovine growth hormone (bGH) has been shown to produce dramatic increases in milk yield of lactating cattle (see Bauman et al., 1985) and as such has provoked considerable and mixed reactions in the dairy industry. Growth hormone also enchances carcass protein : fat ratios in sheep (Muir et al., 1983), pigs (Machlin 1972; Chung et al., 1985; Etherton et al., 1987) and cattle (Brumby, 1959) and may also stimulate mammary development in pubertal heifers (Sejrsen et al., 1986). The nature of these responses was initially considered to be via a lipolytic effect of GH on adipose tissue but it has become apparent that the effects of GH involve a series of changes which result in an altered set-point for energy intake and coordinated changes in different tissues, described as a homeorhetic response (Bauman, 1984).

Despite the tremendous potential for the use of GH (which has led to the production of recombinant human, bovine and porcine hormones) there are still problems involved with its use, most notably regarding administration. Almost all studies have involved daily injection, which is not considered practical, but more recently Rijpkema et al. (1987) demonstrated stimulation of milk yield in dairy cows with fortnightly injections of bovine GH (recombinant bovine somatotrophin, BST). Milk yield did however clearly rise and fall during each 2-week period. Knight et al. (1988) have also described a device which released PST at zero-order rate for 6 weeks in pigs and produced similar effects to daily injections.

MANIPULATION OF ENDOGENOUS GH

As an alternative to exogenous GH, others have attempted to increase endogenous GH either by administration of growth hormone-releasing factor (GRF) or by attempting to neutralize somatostatin secretion (which inhibits GH release). GRF has been shown to increase milk yield in lactating sheep (Hart et al., 1985) and cattle (Lapierre et al., 1988; Enright et al., 1988). Treatment with GRF during pregnancy also increased milk yield in sheep, presumably via a mammogenic effect, possibly involving IGF-I (Kann et al., 1988). Again, because of the peptide nature of GRF, problems of administration must be overcome. Cyclic analogues of GRF (1-29) which appear more stable than the native hormone, show promise in this respect (Campbell et al, 1988).

In addition to GRF, various other secretagogues have been assessed for their ability to stimulate GH release, including adrenergic agents, peptides and amino acids. α-adrenergic agents such as clonidine have been shown to be potent stimulators of growth hormone release, probably acting by increasing GRF release (Miki et al., 1984). Indeed, clonidine treatment of children with constitutional growth delay resulted in increased growth rates over a 1 year period, with increases in both GH and IGF-I evident (Castro-Magaña et al., 1986) although the sedative effects of such centrally-active agents must be cause for concern.

A number of peptides based upon met-enkaphalin have been developed, leading ultimately to a hexapeptide which is a potent GH-releasing agent (Sartor et al., 1985) and which has been shown to increase milk production in lactating dairy cattle (Croom et al., 1984). This peptide does not appear to act through the same mechanism as GRF and the endogenous ligand, which this hexapeptide presumably mimics, has not yet been described.

Intravenous infusion of arginine is a potent stimulus to the secretion of a number of hormones including growth hormone, insulin, prolactin and placental lactogen in a variety of species (see Chew et al., 1984). In a study using lactating dairy cows, intravenous but not abomasal infusion of arginine increased circulating GH concentrations, although the increases were transient and were apparently insufficient to increase milk yield (Vicini et al., 1988). By contrast intravenous arginine administration to pregnant cows increased GH secretion and also tended to increase milk yield (Chew et al., 1984). The quantities of arginine typically required (0.1-0.5 g/kg/d) are large and probably preclude its commercial use.

Successful immunization against somatostatin was first described in St Kilda sheep with a resultant 76% increase in growth rate, although subsequent studies in sheep, cattle, pigs and poultry have suggested more modest increases (See Spencer 1987 for review).

A fascinating observation has also been described whereby monoclonal antibodies when complexed to GH, rather than neutralizing its activity, actually dramatically enhanced its potency (Aston et al., 1986). Several possibilities were proposed for the mechanism of such an effect, including increased half-life, enhanced affinity for GH receptors or selective enhancement of the ability to bind to certain types of GH receptors (presumably those involved with growth). No experimental evidence to support these proposals has been forthcoming and indeed recent studies have shown enhanced activity of GH when conjugated to albumin or indeed to itself (Holder, Morrell and Aston, unpublished observations).

GH FRAGMENTS

A vast effort has gone into the investigation of fragments of GH, in search of a biologically active peptide fragment or "active core" and numerous peptides exhibiting one or other of GH's diverse functions have been described (for review see Paladini et al., 1983) although the physiological significance or commercial use of any has yet to be demonstrated. Interest and concern has been shown about reports that limited tryptic digests of bovine GH produces fragments with metabolic activities similar to human GH (Yamasaki et al., 1970) although with greatly reduced biological activity. Such reports have not however been confirmed using recombinant bovine GH.

INSULIN-LIKE GROWTH FACTOR-I

The mechanism of action of GH in stimulating body growth and milk production is still not clear. Daughaday (1981) proposed that GH works through the stimulation of IGF-I production, which serves as the true growth promoter, whilst others (Nilsson et al., 1987) have shown unilateral growth of long bones in hypophysectomized rats using close arterial infusion of GH, whilst IGF-I failed to evoke a similar response. Paradoxically the effects of close arterial infusion of GH could be prevented by an antiserum to IGF-I suggesting that locally produced IGF-I mediates this effect of GH (Schlechter et al., 1985). At least 2 other groups have also described growth responses to systemic IGF-I in hypophysectomized rats (Hizuka et al., 1987; Zapf et al., 1987). Commercial interest in IGF-I is apparently limited, probably because of its scarcity and because, as a peptide which circulates in high concentrations, the quantities required and mode of administration remain a problem. It is worth considering, however, the potential use of IGF-I in circumstances where the lipolytic or anti-lipogenic effects of GH are not required, since these are thought to be mediated via direct effects of GH on adipose tissue (see Boyd and Bauman 1988, for review).

In addition to the selective effects of IGF-I, it may be possible to overcome the limitations of having to administer large amounts of IGF-I exogenously by enhancing the activity of endogenous IGF-I (rather than increasing its concentration, as GH does). The majority of IGF-I circulates in the blood bound to a 150 Kd binding protein, whilst many cells secrete a lower molecular weight binding protein(s) which inhibits IGF-I activity. The possibility of using antibodies to prevent such interactions could serve to enhance IGF-I bioactivity. In support of this proposal Ballard (1988) has described a truncated form of IGF-I, lacking 3 amino acids at the N-terminus, which fails to bind to the low molecular weight binding protein and shows considerably enhanced biological activity in vitro. In diabetes, malnutrition and uremia, a variety of ill-defined inhibitors of insulin-like growth factors have been described which may also be susceptible to manipulation in order to enhance IGF-I activity (Philips et al., 1984; Taylor et al., 1987). A binding protein for growth hormone in human serum has also been described (Baumann et al., 1986) and although its role in GH action is unclear, it represents another potential point of control.

ANTI-IDIOTYPIC ANTIBODIES AS HORMONE IMAGES

An alternative approach currently under investigation in our own laboratories is the possible use of the anti-idiotypic network of the immune system to manufacture growth hormone "look-alikes". Anti-idiotypes

have been produced which mimic insulin action (Sege and Peterson 1978) and
β adrenergic agents (Schreiber et al., 1980), whilst naturally occurring
autoantibodies with hormonal effects have also been described, such as
long acting thyroid stimulator (LATS), an immunoglobulin which binds to
the TSH receptor (Weetman and McGregor, 1984). We have successfully
produced anti-idiotypic antibodies to rat GH which specifically displace
^{125}I-labelled GH from GH receptors in liver and adipose tissue of both
rats and sheep and which increased body weight gain in hypophysectomized
rats (Flint and Gardner, unpublished observations).

Such anti-idiotypic antibodies would have two potential advantages
over GH itself. Firstly anti-idiotypes could be induced by immunization,
using the idiotypic antibody, which would lead to increased growth hormone
images, possibly for several months at a time. Secondly, it should be
possible to produce monoclonal anti-idiotypes which mimicked only a
portion of the GH molecule. This would permit examination of the
possibility that different parts of the GH molecule possess different
bioactivities or, alternatively that different parts of the GH molecule
interact with different GH receptor subtypes (Barnard et al., 1985; Thomas
et al., 1987).

β-AGONISTS

Perhaps the major alternative to GH under consideration, at least in
terms of carcass composition, is the use of β-agonists. In particular
cimaterol and ractopamine show marked ability to increase lean tissue
whilst decreasing fat deposition in a number of species (for review see
Hanrahan et al., 1986). Based upon the natural hormone adrenaline, these
compounds have become exceptionally interesting because they exert direct
effects on muscle protein synthesis and degradation (Bergen et al., 1987;
Eadara et al., 1987; Maltin et al., 1987). They are attractive from the
point of view that they are orally active and can hence be incorporated
into feedstuffs but this presumably makes their handling more of a problem
(since they are active in humans) and they also suffer from the
requirement for extremely short withdrawal periods before effects are lost
(about 1 week). There is also concern that these agents have deleterious
effects on meat quality (Warriss and Kestin, 1988).

ANTIBODIES TO ADIPOSE TISSUE

A completely different approach to modifying body composition has
been developed in our laboratories. It involves the use of antibodies to

adipocytes, which are capable of destroying such cells in vivo resulting
in considerably reduced body fat deposition in rats (Flint et al., 1986;
Futter and Flint, 1987). In subsequent experiments antibodies to
adipocytes have been shown to increase liveweight gain and food conversion
efficiency, whilst we have also demonstrated long-term localized
reductions of subcutaneous fat in pigs (Kestin, Tonner and Flint,
unpublished observations). Reductions in fat deposition in sheep treated
in similar fashion have also been reported (Moloney and Allen, 1988).

ANTIBODIES TO ENDOCRINE CELLS

We have also used the immune system to selectively ablate pituitary
somatotrophs, leading to long term reductions in body weight gain (Gardner
and Flint, 1988). An alternative approach has been to target toxins
specifically to pituitary cells using the appropriate hypothalamic
releasing factor conjugated to the toxin. Antibodies to somatostatin
could conceivably be used in this way to destroy somatostatin secreting
cells and thereby enhance endogenous GH release.

TRANSGENIC ANIMALS

All of these manipulations are threatened by the proposition of
transgenic animals which express additional GH genes leading to elevated
circulating GH concentrations. First described in giant mice (Palmiter et
al., 1982) and now in pigs (Pursel et al., 1988) it provides a powerful
approach, particularly if the degree of expression can be appropriately
regulated during the animal's lifespan. Such approaches inevitably result
in concern for animal welfare and this particular transgenic approach has
not been without its early problems. It would seem certain that exposure
to very high concentrations of GH over prolonged periods does lead to
changes not only in body composition (ie protein, fat, ash) but also body
proportions. Whether due to GH injections, GH secreting tumours (McCusker
and Campion, 1986) or transgenetics, liver size increases dramatically and
to a much greater extent than total body size. In transgenic mice for
example all organs measured, with the notable exception of the brain, were
increased relative to the increase in body size (Shea et al., 1987).

CONCLUSIONS

In summary, it would seem that bovine and porcine GH represent a
potentially useful and safe way of enhancing both body composition and
milk production. Being protein in nature and inactive in humans, residues

in meat should be of no concern. These effects can be achieved by exogenous GH or endogenous GH evoked by GRF injection. Alternative strategies using the immune system to neutralize somatostatin, or to produce GH mimics, or even to directly destroy adipocytes would appear to present equally safe and acceptable approaches, whilst the potential benefits of the use of IGF-I may lie in more subtle requirements for particular actions of GH.

Although transgenic animals would appear to be the ideal approach, the stability of transgene transmission for commercial species has yet to be evaluated and regulation of gene expression is also essential so that GH secretion can be switched on and off at appropriate times. This approach inevitably suffers somewhat from a lack of flexibility in order to permit decisions "to treat or not to treat" on an individual basis or in response to rapid alterations in consumer demand.

The recent description of the 3-dimensional structure of porcine growth hormone using crystal x-ray diffraction (Abdel-Meguid et al., 1987) and the publication of the amino acid sequences of rabbit and human growth hormone receptors (Leung et al., 1987) are tremendous steps forward towards the definition of the "active core" of the growth hormone molecule and to a greater understanding of the molecular basis of growth hormone action. Such developments may alter radically our approaches to manipulation of animal performance.

REFERENCES

Abdel-Meguid, S.S., Shieh, H.-S., Smith, W.W., Dayringer, H.E., Violand, B.N. and Bentle, L.A. 1987. Three-dimensional structure of a genetically engineered variant of porcine growth hormone. Proc. Natl. Acad. Sci. USA $\underline{84}$, 6434-6437.

Aston, R., Holder, A.T., Preece, M.A. and Ivanyi, J. 1986. Potentation of the somatogenic and lactogenic activity of human growth hormone with monoclonal antibodies. J. Endocr. $\underline{110}$, 381-388.

Ballard, F.J. 1988. Insulin-like growth factor effects on protein metabolism in cultures cells. Biochem. Soc. Trans. (In press)

Barnard, R., Bundesen, P.G., Rylatt, D.B. and Waters, M.J. 1985. Evidence from the use of monoclonal antibody probes for structural heterogeneity of the growth hormone receptor. Biochem. J. $\underline{231}$, 459-468.

Bauman, D.E. 1984. Regulation of nutrient partitioning. In "Herbivore Nutrition in the Subtropics and Tropics", pp 505-524. Eds F.M.C. Gilchrist and R.I. Machie. The Science Press, Craighall, South Africa.

Bauman, D.E., Eppard, P.J., De Geeter, M.J. and Lanza, G.M. 1985. Responses of high-producing dairy cows to long-term treatment with pituitary somatotrapin and recombinant somatotropin. J. Dairy Sci. $\underline{68}$, 1352-1362.

Baumann, G., Stolar, M.W., Amburn, K., Barsano, C.P. and de Vries, B.C. 1986. A specific growth hormone-binding protein in human plasma: Initial characterization. J. Clin. Endocr. Metab., 62 134-141.

Bergen, W.G., Johnson, S.E., Skjaerlund, D.M., Merkel, R.A. and Anderson, D.B. 1987. The effect of ractopamine on skeletal muscle metabolism in pigs. Fed. Proc., 46, 1021.

Boyd, R.D. and Bauman, D.E. (1988) In "Current Concepts of Animal Growth" (Ed. D.R. Campion, G.J. Houseman and R.J. Martin). (Planum Press). (In Press)

Brumby, P.J. 1959. The influence of growth hormone on growth in young cattle. New Zealand J. Agric. Res., 2, 683-686.

Campbell, R.M., Su, C.-M., Maines, S.L., Stricker, P.R., Jensen, L.R., Heimer, E.P., Felix, A.M. and Mowles, T.F. 1988. Biological activities of novel cyclic growth hormone-releasing factor (GRF) analogs. J. Anim. Sci., 66 Suppl 1, 291.

Castro-Magaña, M., Angulo, M., Fuentes, B., Castelar, M.E., Cañas, A. and Espinoza, B. 1986. Effect of prolonged clonidine administration on growth hormone concentrations and rate of linear growth in children with constitutional growth delay. J. Pediatr., 109, 784-787.

Chew, B.P., Eisenman, J.R. and Tanaka, T.S. 1984. Arginine infusion stimulates prolactin, growth hormone, insulin, and subsequent lactation in pregnant dairy cows. J. Dairy Sci., 67, 2507-2518.

Chung, C.S., Etherton, T.D. and Wiggins, J.P. 1985. Stimulation of swine growth by porcine growth hormone. J. Anim. Sci., 60, 118-130.

Croom jr W.J., Leonard, E.S., Baker, P.K., Kraft, L.A. and Ricks, C.A. 1984. The effects of synthetic growth hormone releasing hexapeptide BI 679 on serum growth hormone levels and production in lactating dairy cattle. J. Dairy Sci. (Suppl 1), 67, 109.

Daughaday, W.H. 1981. Growth hormone and the somatomedins. In: "Endocrine control of growth" (Ed W.H. Daughaday) (Elsevier N.Y.) pp 1-24.

Eadara, J., Dalrymple, R.H., De Lay, R.L., Ricks, C.A. and Romsos, D.R. 1987. Cimaterol, a novel β-agonist, selectively stimulates white adipose tissue lipolysis and skeletal muscle lipoprotein lipase activity in rats. Fed. Proc., 46, 1020.

Enright, W.J., Chapin, L.T., Moseley, W.M. and Tucker, H.A. 1988. Effects of infusions of various doses of bovine growth hormone-releasing factor on growth hormone and lactation in Holstein cows. J. Dairy Sci., 71, 99-108.

Etherton, T.D., Wiggins, J.P., Evock, C.M., Chung, C.S., Rebhun, J.F., Walton, P.E. and Steele, N.C. 1987. Stimulation of pig growth performance by porcine growth hormone: determination of the dose-response relationship. J. Anim. Sci., 64, 433-443.

Flint, D.J., Coggrave, H., Futter, C.E., Gardner, M.J. and Clarke, T.J. 1986. Stimulatory and cytotoxic effects of an antiserum to adipocyte plasma membranes on adipose tissue metabolism in vitro and in vivo. Int. J. Obesity 10, 69-77.

Futter, C.E. and Flint, D.J. 1987. Long-term reduction of adiposity in rats after passive immunization with antibodies to rat fat cell plasma membranes. In "Recent Advances in Obesity Research V". Chapter 27. (Ed. E.M. Berry). (John Libbey and Co., London).

Gardner, M.J. and Flint, D.J. 1988. Effects of an antiserum to rat growth hormone (rGH) on growth and serum IGF-I levels in neonatal rats. J. Endocrinol., 117 (Suppl), 55.

Hanrahan, J.P., Quirke, J.F., Bomann, W., Allen, P., McEwan, J.C., Fitzsimons, J.M., Kotzian, J. and Roche, J.F. 1986. β-agonists and their effects on growth and carcass quality. In "Recent Advances in Animal Nutrition" (Ed. W. Haresign and D.J.A. Cole). (Butterworth, London). pp. 125-138.

Hart, I.C., Chadwick, P.M.E., James, S. and Simmonds, A.D. 1985. Effect of intravenous bovine growth hormone or human pancreatic growth hormone-releasing factor on milk production and plasma hormones and metabolites in sheep. J. Endocr., 105, 189-196.

Hizuka, N., Takano, K., Asakawa, K., Miyakawa, M., Tanaka, I., Horikawa, R., Hasegawa, S., Mikasa, Y., Saito, S., Shibasaki, T. and Shizume, K. 1987. In vivo effects of insulin-like growth factor I in rats. Endocrinol. Japon., 34 (Suppl 1), 115-121.

Holder, A.T., Aston, R., Preece, M.A. and Ivanyi, J. 1985. Monoclonal antibody-mediated enhancement of growth hormone activity in vivo. J. Endocrinol., 107, R9-R12.

Kann, G., Périer, A. and Martinet, J. 1988. Use of human growth hormone releasing factor (hGRF 1-29)NH_2 as a mammotropic hormone in the ewe. J. Anim. Sci., 66 (Suppl 1) 389.

Knight, C.D., Azain, M.J., Kasser, T.R., Sabacky, M.J., Baile, C.A., Buonomo, F.C. and McLaughlin, C.L. 1988. Functionality of an implantable 6-week delivery system for porcine Somatotropin (PST) in finishing hogs. J. Anim. Sci., 66 (Suppl 1), 257-258.

Lapierre, H., Pelletier, G., Petitclerc, D., Dubreuil, P., Morisset, J., Gaudreau, P., Couture, Y. and Brazeau, P. 1988. Effect of human growth hormone-releasing factor (1-29)NH_2 on growth hormone release and milk production in dairy cows. J. Dairy Sci., 71, 92-98.

Leung, D.W., Spencer, S.A., Cachianes, G., Hammonds, R.G., Collins, C., Henzel, W.J., Barnard, R., Waters, M.J. and Wood, W.I. 1987. Growth hormone receptor and serum binding protein: purification, cloning and expression. Nature, 330, 537-543.

Machlin, L.J. 1972. Effect of porcine growth hormone on growth and carcass composition of the pig. J. Anim. Sci., 35, 794-800.

Maltin, C.A., Hay, S.M., Delday, M.I., Smith, F.G., Lobley, G.E. and Reeds, P.J. 1987. Clenbuterol, a beta agonist, induces growth in innervated and denervated rat soleus muscle via apparently different mechanisms. Bioscience Reports, 7, 525-532.

McCusker, R.H. and Campion, D.R. 1986. Effect of growth hormone-secreting tumors on body composition and feed intake in young female Wistar-Furth rats. J. Anim. Sci., 63, 1126-1133.

Miki, N., Onu, M. and Shizume, K. 1984. Evidence the opiatergic and α-adrenergic mechanisms stimulate rat growth hormone release via growth hormone releasing factor (GRF). Endocrinology, 114 1950-1952.

Moloney, A.P. and Allen, P. 1988. Growth and weights of abdominal and carcass fat in sheep immunized against adipose cell membranes. Proc. Nutr. Soc., (In press)

Muir, L.A., Wien, S., Duquette, P.F., Rickes, E.L. and Cordes, E.H. 1983. Effects of exogenous growth hormone and diethylstibestrol on growth and carcass composition of growing lambs. J. Anim. Sci., 56, 1315-1323.

Nilsson, A., Isgaard, J., Lindahl, A., Peterson, L. and Isaksson, O. 1987. Effects of unilateral arterial infusion of GH and IFG-I on tibial longitudinal bone growth in hypophysectomized rats. Calcif. Tiss. Int., 40, 91-96.

Paladini, A.C., Peña, C. and Poskus, E. 1983. Molecular biology of growth hormone. CRC Crit. Rev. Biochem., 15, 25-56.

Palmiter, R.D., Brinster, R.L., Hammer, R.E., Trumbauer, M.E., Rosenfeld, M.G., Birnberg, N.C. and Evans, R.M. 1982. Dramatic growth of mice that develop from eggs microinjected with metallothionine-growth hormone fusion genes. Nature, 300, 611-615.

Phillips, L.S., Fusco, A.C., Unterman, T.G. and del Greco, F. 1984. Somatomedin inhibitor in uremia. J. Clin. Endocrinol. Metab., 59, 764-772.

Pursel, V.G., Campbell, R.G., Miller, K.F., Behringer, R.R., Palmiter, R.D. and Brinster, R.L. 1988. Growth potential of transgenic pigs expressing a bovine growth hormone gene. J. Anim. Sci., 66 (Suppl 1), 267.

Rijpkema, Y.S., Reeuwijk, L. van, Peel, C.J. and Mol, E.P. 1987. Responses of dairy cows to long-term treatment with somatotropin in a prolonged release formulation. Eur. Assoc. Anim. Prod., Lisbon, September meeting.

Sartor, O., Bowers, C.Y., Reynolds, G.A. and Momany, F.A. 1985. Variables determining the growth hormone response of His–D–Trp–Ala–Trp–D–Phe–Lys–NH_2 in the rat. Endocrinology, 117, 1441-1447.

Schlechter, N.L., Russell, S.M., Spencer, E.M. 1985. Prog. Endocrine Soc. Annual Mtg. Abs. 254.

Schreiber, A.B., Courgud, P.D., Andre, C.L., Vray, B. and Strosberg, A.D. 1980. Anti-alprenolol anti-idiotypic antibodies bind to β-adrenergic receptors and modulate catecholamine sensitive adenylate cyclase. Proc. Natl. Acad. Sci. USA., 77, 7385-7389.

Sege, K. and Peterson, P.A. 1978. Use of anti-idiotypic antibodies as cell surface receptor probes. Proc. Natl. Acad. Sci. USA, 75, 2443-2447.

Sejrsen, K., Foldager, J., Sorensen, M.T., Akers, R.M. and Bauman, D.E. 1986. Effect of exogenous bovine somatotropin on pubertal mammary development in heifers. J. Dairy Sci., 69, 1528-1535.

Shea, B.T., Hammer, R.E. and Brinster, R.L. 1987. Growth allometry of the organs in giant transgenic mice. Endocrinology, 121, 1924-1930.

Spencer, G.S.G. 1987. Biotechnology in the potential practical application of somatotrophic hormones for improving animal performance. Reprod. Nutr. Dévelop., 27 (2B), 581-589.

Taylor, A.M., Sharma, A.K., Avasthy, N., Duguid, I.G.M., Blanchard, D.S., Thomas, P.K. and Dandona, P. 1987. Inhibition of somatomedin-like activity by serum from streptozotocin-diabetic rats: prevention by insulin treatment and correlation with skeletal growth. Endocrinology, 121, 1360-1365.

Thomas, H., Green, I.C., Wallis, M. and Aston, R. 1987. Heterogeneity of growth-hormone receptors detected with monoclonal antibodies to human growth hormone. Biochem. J., 243, 365-372.

Vicini, J.L., Clark, J.H., Hurley, W.L. and Bahr, J.M. 1988. Effects of abomasal or intravenous administration of arginine on milk production, milk composition, and concentrations of somatotropin and insulin in plasma of dairy cows. J. Dairy Sci., 71, 658-665.

Warriss, P.D. and Kestin, S.C. 1988. Beta-agonists improve the carcass but may reduce meat quality in sheep. Anim. Prod., 46, 502.

Weetman, A.P. and McGregor, A.M. 1984. Autoimmune thyroid disease: developments in our understanding. Endocrine Reviews, 5, 309-355.

Yamasaki, N., Kikutani, M. and Sonenberg, M. 1970. Peptides of a biologically active tryptic digest of bovine growth hormone. Biochemistry, 9, 1107-1114.

Zapf, J., Scheiwiller, E., Guler, H.P. and Froesch, E.R. 1987. Insulin and Insulin-like Growth Factor I: Comparative aspects of their in vivo actions on growth and glucose homeostasis. Endocrinol. Japon., 34 (Suppl 1) 123-129.

LONG-TERM EFFECTS OF RECOMBINANT BOVINE SOMATOTROPIN (rBST) ON DAIRY COW PERFORMANCES : A REVIEW

Y. CHILLIARD

Lactation Laboratory - INRA-Theix
63122 CEYRAT (France)

ABSTRACT

Long-term (18-32 week) effects of recombinant BST (bovine somatotropin) injected daily (650 cows, 20 assays) or as prolonged release preparations (790 cows, injected each 14 or 28 d., 23 assays) are reviewed. Daily injections of 10-15, 20-27 or 31-50 mg BST increased milk yield of 3.9, 5.2 or 5.6 kg/d (respectively). Somidobove (320, 640 or 960 mg BST each 28 d.) increased milk yield 2.2, 3.1 or 4.2 kg/d, and Sometribove (500 mg/14 d.) 4.4 kg/d. Mean milk composition was unchanged, but cycled between injections in the case of prolonged release preparations. Dry matter intake increase for high BST doses was 1.5-1.7 kg/d in 32-week experiments, and 0.8 kg/d in 18 week-experiments. Energy balance and body condition score decreased during the first 3 months of BST treatment, due to delayed DM intake increase. In two experiments body lipids were sharply decreased in BST cows. Feed efficiency was increased (6-17 %) according to dilution of maintenance requirement in total needs and to decreased body tissue deposition. There was no clear interaction of BST response with cow parity and milk yield potential, but a trend towards decreased response after several months of BST. High energy-high protein complete mixed rations tended to increase BST response, but inter-assay variability was high. Responses tended to be lower (2.5 $\pm$ 1.8 kg milk/d) at pasture in European conditions.

Knowledge of the long-term (several months) effects of BST (bovine somatotropin, or growth hormone) in dairy cows has increased rapidly since 1985, thanks to the increasing production of BST by recombinant DNA techniques. Since reviews by Peel et Bauman (1987), Chilliard (1988) and Hart (1988) a great number of new results have appeared, both from America (1988, Am. Dairy. Sci. Assoc. Meeting) and Europe (see below). These results are more or less the first concerning the effects of BST in prolonged release preparations, that allow large scale field experiments in practical conditions.

This paper reviews the effects of long-term BST injections on dairy cow performances : milk production and composition, dry matter intake and energy balance, body weight and body condition changes, feed efficiency,

TABLE 1 - Short-term (5-21 d.) effects of BST on performances of dairy cows with different calculated energy balances (CEB) (a).

CEB of BST cows	Milk yield* kg/d	%	Milk fat* (g/l)	Milk pro-* tein (g/l)	DMI (d)* (kg/d)	CEB*
3.3 (±1.4)(b)	+4.5 (±1.6)	+19.6 (±7.9)	+0.8 (±2.3)	-0.8 (±0.8)	-0.2 (±1.1)	-3.7 (±2.3)
-3.6 (±3.9)(c)	+3.5 (±0.7)	+17.3 (±8.6)	+3.1 (±2.5)	-2.1 (±1.7)	-0.8 (±1.0)	-4.8 (±1.8)

(a) See Chilliard (1988) for references (b) 9 assays, 47 cows, CEB of BST cows > 1.1 Mcal/d.
(c) 9 assays, 46 cows, CEB of BST cows < 0.4 Mcal/d. (d) Dry matter intake (mean ± s.d.).

(*) BST-control.

TABLE 3 - Long-term effects of pituitary BST on cow milk yield.

Reference	N. cows	Dose (mg/d.)	Weeks of BST	Milk yield Control	(kg/d) BST-control
Brumby and Hancock, 1955	2 x 3 (a)	50	12	13.1	5.8
Peel et al., 1985	2 x 5 (a)	50	22	19.8	3.5
Bauman et al., 1985	2 x 6	27	27	27.9	4.6
Hutchison et al., 1986	2 x 6	27	27	26.0	2.0
Mean				21.7	4.0

(a) sets of twin cows

in connection with BST dose levels, BST release preparations, cow parity, milk production potential, lactation stage, feeding and management factors.

SHORT-TERM EXPERIMENTS

Most of these experiments (1973-1986) were done by injecting the natural pituitary somatotropin. The first experiment with recombinant BST was published by Bauman et al. (1982).

During 18 short-term assays (less than 4 weeks, after the second month of lactation) the mean **milk production** reponse to BST (25-50 units/day) was 4.0 (±1.3) kg/day, whereas **dry matter intake** (-0.5 ± 1.1 kg/day) and calculated **energy balance** (-4.2 ± 2.1 Mcal/day) decreased (Chilliard, 1988). Effects on **milk composition** differed according to the energy balance of the treated cows (table 1) in accordance with the well known general effects of energy balance on milk fat and protein contents (Journet et Chilliard, 1985 ; Rémond, 1985). Lactose content tended to increase (+0.6 ± 1.1 g/l) whatever the energy balance of the cows.

The milk response (% or kg/day) was lower during the first two months of lactation than after peak yield (6 vs. 12-30 %) (Chilliard, 1988). Response to increasing doses of BST was curvilinear (Eppard et al., 1985).

LONG-TERM EFFECTS OF RECOMBINANT BST INJECTIONS
A/ MILK PRODUCTION AND COMPOSITION
1- Daily injections

In 21 experiments on 969 dairy cows (mean duration of 32 weeks, beginning 5-13 weeks after calving) there was a dose - dependent response (table 2) : the mean (weighed for cow number) increase in **milk production** was 2.8 kg/day (+11 % over control cows) with 5 mg BST/day, and 5.6 kg/day with 31-50 mg BST/day (+21 % over control cows). Inter-assay S.D. was 2.0 - 2.4 kg/day, i.e. the variation coefficient (%) of the response tended to decrease at higher doses, but was high (39 %) even with 31-50 mg BST/d.. The response to rBST was very rapid (maximum after one week or less). Few long-term assays have been conducted with the pituitary somatotropin (table 3). In two direct comparisons (Bauman et al., 1985 ; Hutchison et al., 1986), the response was lower than with the same dose (27 mg/day) of recombinant methionyl - BST.

The milk fat, protein and lactose contents were unchanged (mean values over the period). When significant changes are reported, they are of limited extent, and positive or negative in different studies.

TABLE 2 - Long-term effects of daily injections of recombinant BST on dairy cow milk yield.

REFERENCE	BST (a)	N. COWS	WEEKS of BST	MILK YIELD (kg/d.) 0	5	BST DOSE (mg/d.) 10-15 BST-CONTROL	20-27	31-50
1 (b)	M	24	27	27.9		6.5	10.1	11.5
2	A	32	38	23.7		3.9	3.2	5.0
3	A	30	38	23.4		4.4	5.3	7.4
4 (b)	?	24	27	26.0		8.0	7.0	5.0
5 (b, e)	A	26	37	29.8		1.8	7.0	10.7
6 (b)	A	38	38	26.7		3.8	4.8	4.1
7 (b)	A	127	?	27.7		3.3	4.4	5.4
8 E	?	80	37	20.1		4.0	4.8	4.7
9 (b)	A	120	29	?		4.0	5.0	5.0
10	A	48	30	23.7		6.6	7.8	6.3
11	A	32	38	21.1	4.4	5.3	8.2	
12	U	32	33	21.0	3.3	2.2	4.4	
13 (d)	?	40	31	21.1		2.3	3.2	4.1
14	A	36	38	30.9		-0.3	1.7	4.8
15 (b)	A	28	38	29.9		3.5	7.6	7.0
16 (b,f)	?	32	33	22.5	0.6	5.5	4.2	
17	A	30	38	27.4			4.5	
17 (e)		30		27.4			2.2	
18 (c)	?	32	20	31.7			4.3	
19 (d)	?	32	31	26.1			6.1	
20 (g)	M	96	12	32.5			6.1	
MEAN			32	26.0	2.8	4.1	5.3	6.2
S.D.			7	3.8	2.0	2.1	2.1	2.4
N. COWS		969		267	24	190	300	158
W.MEAN (h)				27.0	2.8	3.9	5.2	5.6

(a) A = American Cyanamid ; M = Monsanto ; U = Upjohn - (b) 3.5 p.100 FCM - (c) 4 p.100 FCM
(d) 305 d. lactation - (e) 2nd BST consecutive lactation - (f) 50 % of Jersey cows -
(g) milking 3 times/day - (h) weighed mean (for cow number) - E = Europe.

(1) Bauman et al., 1985 ; (2) Baird et al., 1986 ; (3) Chalupa et al., 1986, 1987a ;
(4) Hutchison et al., 1986 ; (5) Annexstad et al., 1987 ; (6) Burton et al., 1987 ;
(7) Chalupa et al., 1987 b ; (8) Thomas et al., 1987 ; (9) Chalupa et al., 1988 ;
(10) Eisenbeisz et al., 1988b ; (11) Elvinger et al., 1988 ; (12) Munneke et al., 1988 ;
(13) Nytes et al., 1988 ; (14) Palmquist, 1988 ; (15) Soderholm et al., 1988 ;
(16) West et al., 1988 ; (17) Hemken et al., 1988 ; (18) Rowe-Bechtel et al., 1988 ;
(19) Tessmann et al., 1988 ; (20) Aguilar et al., 1988.

2- Prolonged release preparations

Different companies are producing somatotropin molecules whose precise structure, as well as excipients used for prolonged release systems, are still proprietary information. Moreover, BST dose and injection frequencies differ between companies. As a consequence, results for each product will be presented separately in this chapter.

Somidobove (Eli Lilly) was used at three doses (**320, 640, 960 mg**) and injected every **28 d.**, corresponding theoretically to 11, 23 and 34 mg rBST/d., respectively. Production responses (weighed means) on about 100 cows per dose (table 4) were 56, 60 and 75 % (respectively) of those obtained by daily injections in comparable amounts (table 2). Increasing the injection frequency (28, 21 and 14 d.) of the same amount of BST did not change milk response to any large extent (Mc Guffey et al., 1987b ; Vérité et al., 1988). Part of the difference with daily injections could also be due to feeding or management factors (see below). Milk fat and protein content were unchanged in most experiments. However increases were observed with 320 mg/28 d. injections (Oldenbroek et al., 1987 ; Vérité et al., 1988 ; Vignon et al., 1988).

Sometribove (Monsanto) was used as a **500 mg/14 d.** formulation, corresponding theoretically to 36 mg rBST/d. Milk yield increase in about 450 cows (4.4 kg/d., + 19 % over control, table 5) was 79 % of that obtained by daily injections (table 2). Higher responses (8-13 kg/d.) have been obtained with 1800 or 3000 mg/14 d. (Eppard et al., 1988). Milk composition was unchanged in most experiments. However milk protein content was significantly increased (+0.9 to 1.0 g/l) in 4 experiments (Phipps, 1987 ; Bauman et al., 1988 ; Samuels et al., 1988 ; White et al., 1988). This increase was observed during the end of lactation (Phipps, 1987) and it would be suitable to weigh it for decreasing milk volumes to evaluate its quantitative significance.

A prolonged release preparation from **American Cyanamid** was also experimented (Jenny et al., 1988) and gave 3.5 % FCM responses of 4.9 - 4.0 and 4.4 kg/d for injections of 140-350 and 700 mg each 14 d. during 26 weeks (9 cows/group). Milk production responses to different injection strategies are summarized in table 6.

Apart from the mean production and composition, it is of interest for milk recording and milk quality controls to know if there are some **fluctuations during the period between two BST injections**. The milk response to BST was indeed time-dependent (Huber, 1987 ; Oldenbroek et

TABLE 4 — Long-term effects of Somidobove injections on dairy cow milk yield

REFERENCE	N. COWS	WEEKS of BST	MILK YIELD (kg/d.) BST Dose (mg/28 d.)			
			0	320	640 BST- Control	960
21	14	12	22.9			4.1
22 (a)	70	12	26.4	3.9	4.9	6.3
23 E (b)	48	24	20.9	0.9	3.3	2.9
24	16	16	29.8		2.4	
25 (c)	188	12-36	26.8	2.7	4.3	4.8
26 E	36	24	24.8	0.2	2.7	2.2
27 E	24	20	21.0		1.2	
28 E	40	12	16.3	-0.3	0.1	0.4
29 E (d)	32	20	17.7		1.3	
MEAN		18	23.0	1.5	2.5	3.4
S.D.		5	4.4	1.8	1.6	2.1
N. COWS	468		129	99	136	104
W. MEAN (e)			23.6	2.2	3.1	4.2

(a) Mean of 14, 21 and 28 d. injection frequencies - (b) mixed breeds -
(c) pooled data from 5 herds - (d) Montbéliarde cows - (e) weighed mean.

(21) Mc Guffey et al., 1987a ; (22) Mc Guffey et al., 1987b ; (23) Oldenbroek et al. 1987 ;
(24) Mc Guffey et al., 1988 ; (25) Meyer et al., 1988 ; (26, 27) Vérité et al., 1988 ;
(28) Vignon et al., 1988 ; (29) Parrassin and Vignon, 1988 ; E = Europe.

al., 1987). Maximal response was observed about one week after each injection (Phipps, 1987 ; Lamb et al., 1988 ; Rémond et al., 1988). Negative responses can be obtained during the 4th week in the case of 28 d. injection cycles (Vérité et al., 1988).

Cyclic responses in milk fat and/or protein contents between 14 d. or 28 d. injections have also been observed (Phipps, 1987 ; Oldenbroek et al., 1987; Barbano et al., 1988 ; Bauman et al., 1988 ; Lamb et al., 19888 ; Rémond et al., 1988 ; Vérité et al., 1988), except by White et al. (1988). The cyclic pattern is still not well known and varied between studies. Milk protein content was generally lower after each injection and higher (up to 2 g/l) just before the following injection (Barbano et al., 1988 ; Rémond et al., 1988 ; Vérité et al., 1988). Milk fat content can increase (up to 5 g/l) in parallel with milk yield (Lamb et al., 1988 ; Vérité et al., 1988), probably due to increased body lipid mobilization when yield is maximal (see below). Long chain fatty acid content of milk fat also cycled between two injections (Farries and Profittlich, 1987 ; Lynch et al., 1988 ; Vérité et al., 1988). There was (Rémond et al., 1988) or was not (Bauman et al., 1988) cyclic change in lactose content.

B/ DRY MATTER INTAKE (DMI) AND CALCULATED ENERGY BALANCE (CEB)

In 14 daily-injection experiments, **DMI increased** from 0.6 kg/day for 10-15 mg BST/d, to 1.7 kg/day for 31-50 mg BST/d. The response variability was higher than 45 %. In 6 Somidobove experiments, DMI increased from 0.2 to 0.8 kg/d., depending on the injected dose. Response in 9 Sometribove experiments was 1.5 kg/d. (table 7). DMI was not increased in one experiment with American Cyanamid preparations (Jenny et al., 1988).

Increase of DM intake in BST cows (20-50 mg/d and Sometribove) was exactly that predicted in untreated cows whose milk yield increased to the same extent (Figure 1). This is in accordance with the view that BST cows are in some ways similar to genetically higher yielding cows (Peel and Bauman, 1987 ; Chalupa and Galligan, 1988).

When not given by authors, EB (table 8) was calculated on a net energy basis from DMI, energy concentration of the diet, and 3.5 or 4 % fat corrected milk yield. These calculations should be considered cautiously, particularly when energy content of the diet was not given by authors in abstracts, and roughly estimated (1.5 - 1.65 Mcal/kg DM) according to diet composition.

TABLE 5 - Long-term effects of Sometribove (500 mg BST/14 d.) on dairy cow milk yield.

REFERENCE	N. COWS	WEEKS of BST	MILK YIELD (kg/d.)	
			Control	BST-Control
30 E (a)	90	33	19.2	3.7
31 E	64	33	24.9**	4.6
32	80	36	27.3*	3.1
33 (b)	40	?	? *	8.3
34 (b, c)	18	?	23.1*	6.9
35 E	60	33	21.7**	3.6
36	80	36	27.3	2.3
37	72	36	25.8	3.8
38 (d)	44	34	16.8*	5.6
39 E	48	33	19.2	1.8
40	126	36	23.8*	5.2
41 E	58	33	21.9**	3.9
42	42	24	28.1*	8.2
42 (e)	21	24	- *	6.6
43 E (f)	38	30	16.0	3.1
MEAN		32	22.7	4.7
S.D.		4	4.0	2.0
N. COWS	881		410	451
W. MEAN (g)			23.2	4.4

(a) Friesians - (b) 600 mg/14 d. - (c) 2nd consecutive lactation - (d) Jerseys -
(e) IM injections - (f) Normandes - (g) weighed mean ; * 3.5 % FCM ; ** 4 % FCM
(30) Phipps, 1987 ; (31) Rijpkema et al., 1987, 1988 ; (32) Bauman et al., 1988 ;
(33, 34) Eppard et al., 1988 ; (35) Gravert et al., 1988 ; (36) Huber et al., 1988 ;
(37) Lamb et al., 1988 ; Anderson et al., 1988 ; (38) Pell et al., 1988 ;
(39) Rémond et al., 1988 ; (40) Samuels et al., 1988 ; (41) Vedeau et Schockmel, 1988 ;
(42) White et al., 1988 ; Lanza et al., 1988a ; (43) Lossouarn, 1988. E = Europe.

TABLE 6 - Effect of dose and frequency of BST injections on milk yield (kg/d).

Dose (mg)	10-15	20-27	31-50	320 (a)	640 (a)	960 (a)	500 (b)
Frequency (days)	1	1	1	28	28	28	14
Mean	4.1	5.3	6.2	1.5	2.5	3.4	4.7
S.D.	2.1	2.1	2.4	1.8	1.6	2.1	2.0
N. cows	190	300	158	99	136	104	451
W. Mean	3.9	5.2	5.6	2.2	3.1	4.2	4.4

(a) Somidobove ; (b) Sometribove ; (c) see table 2, 4, 5, for references.

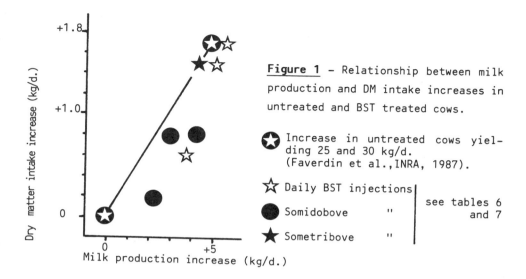

Figure 1 - Relationship between milk production and DM intake increases in untreated and BST treated cows.

⭐ Increase in untreated cows yielding 25 and 30 kg/d. (Faverdin et al., INRA, 1987).
☆ Daily BST injections
● Somidobove   "
★ Sometribove   "
see tables 6 and 7

Daily BST injections **decreased CEB** from 1.2 to 1.8 Mcal/d (table 8). This decrease was smaller for medium dose BST injections where DMI increase was the higher relatively to milk yield increase (tables 6 and 7). In Somidobove experiments CEB decreased 1.4 - 1.9 Mcal/d., whereas it was unchanged in Sometribove experiments. This result could appear surprising since energy needs for 4.4 kg extra-milk are higher than 3 Mcal, which would suppose an energy value higher than 2 Mcal N.E./kg extra-DMI (cf. table 7). This is in fact not surprising if the energy densities of the diets differ according to the milk production level of the cows. Indeed, if 20 kg DMI x 1.59 Mcal/kg = 31.8 Mcal, and 21.5 kg DMI x 1.62 Mcal/kg = 34.8 Mcal, the energy value of the extra 1.5 kg DM would be 2 Mcal/kg. On the other hand, EB calculations are somewhat imprecise due to the limitations pointed out above, and possibly to different methods used by authors in estimating the energy density of the diets.

The lower response in DM intake with Somidobove is probably due in part to the experimental periods (18 ± 5 weeks, table 4) which are shorter than in daily-injection (32 ± 7, table 2) or Sometribove (32 ± 4, table 5) experiments. Indeed, DM intake increase was always delayed, and did not become significant before 6-8 weeks of BST treatment (Bauman et al., 1985; Phipps, 1987 ; Anderson et al., 1988 ; Eppard et al., 1988 ; Dhiman et al., 1988 ; Lanza et al., 1988a). As a consequence, energy balance of

TABLE 7 - Effect of dose and frequency of BST injections on DM intake (kg/d.)

Dose (mg)	10-15	20-27	31-50	320 (a)	640 (a)	960 (a)	500 (b)
Frequency (days)	1	1	1	28	28	28	14
Mean	+0.6	+1.7	+1.7	+0.5	+0.7	+0.7	+1.5
S.D.	0.7	0.9	0.8	0.5	0.4	0.4	0.5
N. cows	116	166	92	89	98	94	337
W. Mean	+0.6	+1.5	+1.7	+0.2	+0.8	+ 0.8	+1.5
References (c)	(1-3, 5-8 , 10-11, 14-18)			(	21-26	)	(d)

(a) Somidobove ; (b) Sometribove ; (c) see table 2, 4, 5 ; (d) 30-32, 35-36, 40-41.

TABLE 8 - Effect of dose and frequency of BST injections on calculated energy balance (Mcal/d.)

Dose (mg)	10-15	20-27	31-50	320 (a)	640 (a)	960 (a)	500 (b)
Frequency (days)	1	1	1	28	28	28	14
Mean	-1.9	-1.2	-2.0	-0.8	-1.2	-1.7	-0.2
S.D.	1.3	1.6	1.1	1.2	0.8	1.0	0.7
N. cows	122	172	98	89	98	94	388
W. Mean	-1.8	-1.2	-1.5	-1.4	-1.4	-1.9	0.0
References (c)	(	1-8, 10-11, 14-18	)	(	21 - 26	)	(d)

(a) Somidobove ; (b) Sometribove ; (c) see table 2, 4, 5 ; (d) 30-32, 34-38, 40-42.

TABLE 9 - Milk yield and calculated energy balance in 8 Sometribove experiments, during the first 12 weeks and the last 22 weeks of BST treatment (315 treated cows) (a).

Period	Milk yield (kg/d.)		Calculated energy balance*	
	Control	BST-Control	Mcal/d	Mcal/period
First 12 weeks	29.5	4.6	-2.1	-176
	(+2.4)	(+1.0)	(+0.7)	
Last 22 weeks	21.0	3.4	+1.2	+185
	(+3.5)	(+1.1)	(+1.0)	

(a) Data from Peel et al. (1988) and references 30-32 , 35-37, 40-41 of table 5.
* BST-Control - The same diet was used throughout each 34-week assay.

treated cows was lower than that of control cows during the first 2 or 3 months of BST administration (Bauman et al., 1985 ; Phipps, 1987 ; Anderson et al., 1988 ; Dhiman et al., 1988 ; Eppard et al., 1988 ; Huber et al., 1988 ; Lanza et al., 1988 a ; Rémond et al., 1988 ; Samuels et al., 1988).

From Sometribove data summarized by Peel et al. (1988a), it was possible to compute EB of the cows during the first 12 weeks and the last 22 weeks of eight 34-week experiments (table 9). CEB of treated cows was higher during the last 22 weeks (due to higher DM intake), and this can compensate completely for lower CEB during the first 12 weeks. However, delayed conception and increasing intercalving intervals (6-21 days in trials beginning during week 9 of lactation) were consequences of this lower CEB during the third month of lactation (see Hard et al., 1988 ; Peel, 1988 and Peel et al., 1988a).

C/ BODY WEIGHT, AND BODY RESERVES OF THE COWS

In high yielding cows, body reserves (mainly lipids) are mobilized during the first 2 months of lactation, and deposited again during decreasing lactation (see Chilliard, 1987, for review).

Few data are available on **body weight** and **body condition score** changes in BST experiments, so results from tables 10 and 11 have to be regarded cautiously. Body weight and body condition change data are however roughly in accordance with **energy balance** data, if we take into account that experimental periods were shorter for Somidobove treatment (daily CEB x day number).

During Sometribove experiments, some authors also observed a lower body condition in BST cows during the first period of treatment, this difference being partly reversed during the following period, in accordance with CEB data (Huber et al., 1988 ; Samuels et al., 1988). This was however not observed in another trial in European conditions (-0.4 and -0.6 points of body condition score during winter and grazing periods, respectively) (Rémond et al., 1988). A lower body condition of BST treated cows (-0.5 point) can also be reversed during the dry period before the next calving (Phipps, 1987).

TABLE 10 - Effect of dose and frequency of BST injections on body weight change (kg) (d).

Dose (mg)	10-15	20-27	31-50	320 (a)	640 (a)	960 (a)	500 (b)
Frequency (days)	1	1	1	28	28	28	14
Mean	-14	-23	-43	0	-5	-18	NS
S.D.	36	33	17	14	13	10	-
N. cows	55	92	38	31	68	29	219
W. Mean	-17	-23	-45	-1	-4	-16	NS
References (c)	(1, 3, 6, 8, 11, 14-16, 18)			(23, 26-29)			(e)

(a) Somidobove ; (b) Sometribove ; (c) see table 2, 4, 5 ; (d) BW change in BST cows - BW change in control cows ; (e) references 30, 32-34, 36, 38-39, 43 ; NS = no significant change.

TABLE 11 - Effect of dose and frequency of BST injections on body condition score (d).

Dose (mg)	10-15	20-27	31-50	320 (a)	640 (a)	960 (a)	500 (b)
Frequency (days)	1	1	1	28	28	28	14
Mean	-1.0	-0.8	-1.7	-0.1	-0.2	-0.4	-0.5
S.D.	0.3	0.5	-	-	-	-	0.4
N. cows	15	95	7	9	22	7	297
W. Mean	-1.0	-0.5	-1.7	-0.1	-0.1	-0.4	-0.2
References (c)	(15, 16, 18-20)			(26, 27 (e))			(f)

(a) Somidobove ; (b) Sometribove ; (c) see table 2, 4, 5 ; (d) 0-5 scale ;
(e) this trend was confirmed by Mc Guffey et al. (1988) and Meyer et al. (1988) ;
(f) references 30, 36, 39, 40.

Body composition and body lipids have been measured using deuteriated water dilution technique in three BST experiments (table 12). In the two experiments exceding 6 months, **body lipid deposition was 16 to 69 kg lower** in BST treated cows whose milk production increased. **Higher body fat mobilization** (blood free fatty acids) was also observed. In another Somidobove experiment (Mc Guffey et al., 1988), subcutaneous fat depth was decreased. Theoretical calculation shows that a 1 Mcal/d. decrease of energy balance (net energy) is equivalent to a 28 kg lipid deposition decrease over 38 weeks (assuming the same efficiency of metabolisable energy for milk secretion and body lipid deposition).

Increases in **blood free fatty acids** were also observed in some BST experiments (Rowe-Bechtel et al., 1988 ; Rémond et al., 1988), but not in others (Oldenbroek et al., 1987 ; Hutchison et al., 1986). Transient increases were observed during the first months of BST when the energy balance of the cows decreased, but not during the remainder of the lactation when DM intake increased (Lanza et al., 1988 b, c).

Blood free fatty acids vary with feeding time. They can also **change during the cycle between two BST injections** (Vérité et al., 1988) as well as long chain fatty acid content of milk fat (Farries and Profittlich, 1987 ; Lynch et al., 1988 ; Vérité et al., 1988), indicating higher body fat mobilization immediately after each BST injection. BST can indeed increase the lipolytic response in cows with low energy balances (Vernon, 1986 ; Chilliard, 1987, 1988 ; Sechen et al., 1988).

Cyclic variations of body weight ($\pm$ 5-10 kg) were observed between Sometribove injections (Phipps, 1987 ; Anderson et al., 1988), but not for DM intake (Bauman et al., 1988). It is not known if this can reflect changes in gut or mammary contents, or in body composition.

D/ FEED EFFICIENCY

Feed efficiency was increased 17 % (over control) by injecting daily 31-50 mg of BST. The corresponding values were 8 % for Somidobove and 6 % for Sometribove (table 13).

Digestibility of the diet components, maintenance requirement and efficiency of metabolisable energy for milk secretion were not significantly increased in BST short-term experiments (Tyrrell et al., 1982 ; Eisemann et al., 1986), nor DM digestibility in long-term experiments (Peel et al., 1985 ; Dhiman et al., 1988 ; Rémond et al., 1988).

TABLE 12 - Effects of BST on body composition of dairy cows.

BST-Control :	Daily injections (a)			Somidobove (b)		Daily injections (c)
	10 mg	21 mg	41 mg	320 mg 28 d.	960 mg 28 d.	40 mg
Milk yield (kg/d)	+3.5	+7.6	+7.0	+0.2	+2.2	Increased
Energy balance (Mcal/d)	-0.9	-1.4	-2.7	+0.2	-1.0	Decreased
Body weight change (kg)	-42	-45	-66	+18	-25	+2
Body lipid change (kg) (d)	-16	-69	-50	+1	-42	-4
Body condition score change(e)	-1.2	-1.6	-1.7	+0.2	-0.2	-
Blood free fatty acids (% over control)	+36	+29	+54	+19	+97	-

(a) Soderholm et al., 1988 : 20 cows, BST over 38 weeks (15 cows).
(b) Vérité and Chilliard, unpublished : 21 cows, BST over 24 weeks (13 cows). Deuteriated water dilution technique was previously standardized on 20 slaughtered dairy cows (Chilliard and Robelin, 1983).
(c) Brown et al., 1988 : 16 cows, BST over 7 weeks (9 cows).
(d) estimated by deuteriated water dilution technique. (e) scale 0-5.

TABLE 13 - Effects of BST on feed efficiency in dairy cows.

DAILY INJECTIONS (mg/d.)		0	10-15	20-27	31-50
FCM/DMI (a)	W. Mean	1.31	1.46	1.47	1.53
	N. Cows	115	108	138	84
FCM/NEI (b)	W. Mean	0.78	0.86	0.89	0.91
	N. Cows	60	49	80	41
SOMIDOBOVE (mg/28 d.)		0	320	640	960
FCM/NEI (c)	W. Mean	0.76	0.81	0.82	0.82
	N. Cows	77	68	77	77
SOMETRIBOVE (mg/14 d.)		0			500
FCM/NEI (d)	W. Mean	0.78			0.83
	N. Cows	367			408

(a) (Fat-corrected) milk/Dry matter intake. References 1,3, 5-7, 10-11, 14-17 in table 2.
(b) (Fat-corrected) milk/Net energy intake. References 1, 3, 10-11, 14-15, 17 in table 2.
(c) References 23-26 in table 4 - (d) References 30-38, 39-42 in table 5.

Increase in feed efficiency resulted 1/ from dilution of maintenance requirement in the total requirement, due to milk yield increase, and 2/ from lower use of nutrients for body tissue deposition, relatively to milk secretion. These two reasons can explain the higher increase of feed efficiency with daily-injection experiments (table 13), since milk response and energy balance decrease were more pronounced than with prolonged release preparations.

E/ INTERACTIONS OF BST RESPONSE WITH ANIMAL AND MANAGEMENT FACTORS

1- Parity :

Lower responses in heifers than in multiparous cows have been reported (Marsh et al., 1987 ; Huber et al., 1988 ; Rowe-Bechtel et al., 1988 ; Tessmann et al., 1988 ; Vérité et al., 1988 ; Whitaker et al., 1988). However there was no difference (or higher responses in heifers) in other studies (Chalupa et al., 1988 ; Hard et al., 1988 ; Lamb et al., 1988 ; Pell et al., 1988 ; Rémond et al., 1988 ; Samuels et al., 1988).

2- Milk production potential :

Most authors observed great individual variations in the response (kg/d.) to BST. It is not known if this variability is repeatable from one lactation to the next. It was not related to individual variations in milk potential (Lanza et al., 1988a ; Sullivan et al., 1988) nor to predicted differences of the sires (Mc Daniel and Hayes, 1988). Negative relationships with individual milk potential were however recorded by Leicht et al. (1987) and Oldenbroek et al.(1987). There was no interassay relationship between milk production of control groups and response to BST in data from tables 2, 4 and 5.

These observations on parity and milk potential effects show that percentage milk response to BST (relative to control) decreases when milk yield increases.

3- Lactation stage and consecutive lactations :

Response to BST did not seem to be dependent on the stage of lactation (after the first month) at which treatment began (Meyer et al., 1988). For a given initial stage, the persistency of the response during the remainder of the lactation was rather variable, but tended often to decrease (Furniss et al., 1988 ; Huber et al., 1988 ; Lossouarn, 1988 ; Phipps, 1987 ; Peel et al., 1988, table 9 ; Rémond et al., 1988 ; Vérité et al., 1988), perhaps in relation to management factors, pregnancy stage, and the physiological need of the cow to maintain body reserves.

TABLE 14 - Comparisons of different diets in long-term BST experiments.

Reference	Dose (mg)	N. Cows per group	Weeks of BST	Milk response (kg/d) "Low" Diet (LD)	"High" Diet (HD)
A	25/d.	10	37	4.6*	4.9
B	21/d.	6	30	7.8	7.8
C	21/d.	10	38	4.5	4.5
D	640/28 d.	8	16	2.4	5.3
E	500/14 d.	12	12	2.3	2.3
F	21/d.	16	31	4.0	6.1

A - Thomas et al., 1987 - LD = flat rate concentrate (9 kg/d.) over 24 weeks, then pasture.
* Lower weight gain over 24 weeks - HD = complete mixed rations.
B - Eisenbeiz et al., 1988 a. Corn vs. barley concentrate.
C - Hemken et al., 1988. 40 % vs. 60 % concentrate.
D - Mc Guffey et al., 1988 ; LD = 14 % protein, 30 % undegradable protein - HD = 17 % protein, 40 % undegradable protein.
E - Rémond et al., 1988. 3.2 vs 4.9 kg/d concentrate (corn silage ad libitum, no difference in total DM intake).
F - Tessmann et al., 1988 ; 12-32 % vs 32-52 % concentrate (alfalfa silage diet).

TABLE 15 - Summary of BST production responses according to feeding diets.

Diet	1 CMD	2-4 CMD	F + C	Pasture
Daily injections (20-27 mg/d.)	4.7 ($\pm$1.8) n = 11 (a) E = 1	6.6 ($\pm$2.2) n = 7 (b)		
Sometribove (500 mg/14 d.)		5.3 ($\pm$2.0) n = 10 (c) E = 2		
Somidobove or Sometribove			3.1 ($\pm$1.3) n = 7 (d) E = 7 4.1 ($\pm$0.8) n = 3 (e) E = 3	2.5 ($\pm$1.8) n = 8 (f) E = 6

1 CMD = one complete mixed ration.
2-4 CMD = 2-4 complete mixed ration of different energent content, according to milk yields.
F + C = (ad libitum forages) + (concentrates) - E = European experiments.
(a) References 2-5, 8, 14-18 (table 2) - (b) References 1,7, 10-11, 13, 19-20 (table 2) -
(c) References 30-34, 36-37, 40, 42 (table 5) - (d) Concentrates according to milk yield - References 26-27, 29, 35, 39, 41, 43 (table 4 and 5) -
(e) Flat rate concentrates - Thomas et al., 1987 ; Furniss et al., 1988 ; Whitaker et al., 1988 - (f) References 27 - 29 (table 4), 39, 43 (table 5), Brumby et Hancock, 1955 ; Peel et al., 1985, 1988 ; Furniss et al., 1988.

BST use during one lactation did not appear to influence yield at the beginning of the subsequent lactation (Phipps, 1987). The percentage of twin calves was increased in BST cows during two European experiments (Phipps, 1987 ; Rijpkema et al., 1987). These observations need to be confirmed. On the other hand, milk responses to BST during a second consecutive lactation were of the same magnitude as during the first (Annexstad et al., 1987 ; Hemken et al., 1988 ; Eppard et al., 1988).

4- <u>Nutrition</u> :

During **short-term experiments**, milk response to BST was not increased by post-ruminal infusions of glucose and casein (Peel et al., 1982), nor by the addition of sodium bicarbonate or branched-chain volatile fatty acids to the diet (Kik and Cook, 1986 ; Chalupa et al., 1984). Although a positive interaction was observed between BST and calcium soaps of fatty acids (Schneider et al., 1987) it was not the case with a hydrolysed blend of fats (Lough et al., 1988).

During **long-term experiments** there was no clear effect of concentrate level or cereal nature (table 14), except in Tessmann et al. (1988). In this experiment, cows receiving low concentrate diet were in lower body condition at the end of BST treatment, and they had a lower persistency during the next BST lactation.

An effect of protein content and degradability was observed by Mc Guffey et al. (1988). The lower response in low protein diet was due to a lower persistency of the response during BST treatment, and the authors suggested it could be due to exhaustion of body protein reserves. Increasing the percentage of undegradable protein did not however increase the response to BST to the same extent in cows of lower milk potential (Vérité et al., 1988).

Data from tables 2, 4 and 5 have been pooled according to feeding diets (table 15). The significance of the different means is poor, since all other experimental conditions were not comparable between assays. It seems however that higher responses could be obtained with several complete mixed rations of different energy content, given according to milk yield. Responses were lower in European feeding conditions. During the 3 winter flat rate concentrate feeding experiments, there was a lower body weight gain in BST cows, but milk response was rather high. The low responses at pasture (table 16) could be due to higher lactation-pregnancy stages (Vérité et al., 1988) or to low quality pasture during drought periods (Lossouarn et al., 1988).

TABLE 16 - Milk response (kg/d) to BST in Europe according to season* and feeding conditions.

Season Diet	Winter Silage + concentrate	Spring and /or summer Pasture (+ Concentrate)	Pasture quality
Furniss et al., 1988 Peel et al., 1988	4.8	4.5	Good
Lossouarn, 1988	5.6	2.8	Low
Parrassin and Vignon, 1988	1.3	1.3	Medium
Rémond et al., 1988	2.3	1.8	Medium
Vérité et al., 1988	2.0	0.3	Good

* Including lactation stage or treatment duration effects.

5- Environmental conditions :

Mollett et al. (1986) observed low response (+2 to +4 %) to 27-40 mg BST/d. over 27 weeks in 18 cows, and hypothesized that hot and humid climatic conditions may have affected treatment response. However Collier and Johnson (1988) summarized data from short- and long-term experiments in cold and hot conditions, and concluded that BST response was not impaired if feeding and management conditions were adequate.

CONCLUSIONS

BST has a great potential for increasing cow milk yield (about +5 kg/d or +1000 kg per lactation for a 200-day treatment with slow release preparations). However, responses can be much lower (1-4 kg/d.), especially in European management conditions. So, percentage increase of lactation yield can vary from less than 5 to more than 25 %, depending on milk production level of untreated cows, on treatment duration and on mean response to BST.

BST decreases energy balance and body reserve gain during the first 3 months of treatment. This leads to increased delay in conception and inter-calving interval, as in higher yielding untreated cows. Cycling of milk composition between injections will lead to increased milk recording and quality control frequencies, but this should be avoided by improving slow release delivery systems in the future.

Increase in feed efficiency (FCM/NEI) results simply from increased milk yield (decreasing the part of maintenance in total needs) and from decreased body tissue gain. If body gain over the period is similar to that of control cows, increase of feed efficiency will be of about 6 % for a 4 kg/d increase in milk yield.

Heifer response was lower in some experiments, but not in others. BST responses were highly variable between adult cows, but not related to individual milk potential. New knowledge is needed to explain this and to improve strategies for BST use. Feeding management is important for milk response to BST, as well as for improving body condition and reproductive performances of the BST treated cows. Present data indicate a lower response (2.5 $\pm$ 1.8 kg/d) to BST in pasture European conditions.

Aknowledgments
I thank Monsanto and Elanco Companies, and particularly C.J. PEEL, for access to unpublished data, and Y. FOURNIER and L. SOUCHET for typing the manuscript.

REFERENCES

AGUILAR A.A., JORDAN D.C., OLSON J.D., BAILEY C., HARTNELL G., 1988. A short- study evaluating the galactopoietic effects of the administration of Sometribove (recombinant methionyl bovine somatotropin) in high producing dairy cows milked three times per day. J. Dairy Sci., 71 (suppl.1), 208 (Abstract).

ANDERSON M.J., LAMB R.C., ARAMBEL M.J., BOMAN R.L., HARD D.L., KUNG L., 1988. Evaluation of a prolonged release system of sometribove, USAN (recombinant methionyl bovine somatotropin) on feed intake, body weight, efficiency and energy balance of lactating cows. J. Dairy Sci., 71 (suppl.1), 208 (Abstract).

ANNEXSTAD R.J., OTTERBY D.E., LINN J.G., HANSEN W.P., SODERHOLM C.G., EGGERT R.G., 1987. Responses of cows to daily injections of recombinant bovine somatotropin (BST) during a second consecutive lactation. J. Dairy Sci., 70 (suppl.1), 176 (Abstract).

BAIRD L.S., HEMKEN R.W., HARMON R.J., EGGERT R.G., 1986. Response of lactating dairy cows to recombinant bovine growth hormone (rbGH). J. Dairy Sci., 69 (suppl.1), 118 (Abstract).

BARBANO D.M., LYNCH J.M., BAUMAN D.E., HARTNELL G.F., 1988. Influence of sometribove (recombinant methionyl bovine somatotropin) on general milk composition. J. Dairy Sci., 71 (suppl.1), 101 (Abstract).

BAUMAN D.E., DeGEETER M.J., PEEL C.J., LANZA G.M., GOREWIT R.C., HAMMOND R.W., 1982. Effects of recombinantly derived bovine growth hormone (bGH) on lactational performance of high yielding dairy cows. J. Dairy Sci., 65 (Suppl.1), 121. (Abstract).

BAUMAN D.E., EPPARD P.J., DeGEETER M.J., LANZA G.M., 1985. Responses of high-producing dairy cows to long-term treament with pituitary somatotropin and recombinant somatotropin. J. Dairy Sci., 68, 1352-1362.

BAUMAN D.E., HARD D.L., CROOKER B.A., ERB H.N., SANDLES L.D., 1988. Lactational performance of dairy cows treated with a prolonged-release formulation of methionyl bovine somatotropin (sometribove). J. Dairy Sci., 71 (suppl.1), 205 (Abstract).

BROWN D.L., TAYLOR S.J., De PETERS E.J., BALDWIN R.L., 1988. Influence of sometribove (a methionyl bovine somatotropin) on the body composition of lactating Holstein cattle. J. Dairy Sci., 71 (suppl.1), 125 (Abstract).

BRUMBY P.J., HANCOCK J., 1955. The galactopoietic role of growth hormone in dairy cattle. N.Z. J. Tech., 36, 417-436.

BURTON J.H., McBRIDE B.W., BATEMAN K., MACLEOD G.K., EGGERT R.G., 1987. Recombinant bovine somatotropin: effects on production and reproduction in lactating cows. J. Dairy Sci., 70 (suppl.1), 175 (Abstract).

CHALUPA W., BAIRD L., SODERHOLM C., PALMQUIST D.L., HEMKEN R., OTTERBY D., ANNEXSTAD R., VECCHIARELLI B., HARMON R., SINHA A., LINN J., HANSEN W., EHLE F., SCHNEIDER P., EGGERT R., 1987b. Responses of dairy cows to somatotropin. J. Dairy Sci., 70 (suppl.1), 176 (Abstract).

CHALUPA W., GALLIGAN D.T., 1988. Nutritional implications of somatotropin for lactating cows. J. Dairy Sci., 71 (suppl.1), 123 (Abstract).

CHALUPA W., HAUSMAN B., KRONFELD D.S., KENSINGER R.S., McCARTHY R.D., ROCK D.W., 1984a. Responses of lactating cows to exogenous growth hormone and dietary sodium bicarbonate. I. Production. J. Dairy Sci., 67 (suppl.1), 107 (Abstract).

CHALUPA W., KUTCHES A., SWAGER D., LEHENBAUER T., VECCHIARELLI B., SHAVER R., ROBB E., 1988. Responses of cows in a commercial dairy to somatotropin. J. Dairy Sci., 71 (suppl.1), 210 (Abstract).

CHALUPA W., MARSH W.E., GALLIGAN D.T., 1987a. Bovine somatotropin:lactational responses and impacts on feeding programs. Proc. Maryland. Nutr. Conf. Feed Manuf., 48-57.

CHALUPA W., VECCHIARELLI B., SCHNEIDER P., EGGERT R.G., 1986. Long-term responses of lactating cows to daily injection of recombinant somatotropin. J. Dairy Sci., 69 (suppl.1), 151 (Abstract).

CHILLIARD Y., 1987. Revue bibliographique: Variations quantitatives et métabolisme des lipides dans les tissus adipeux et le foie au cours du cycle gestation-lactation. 2ème partie: chez la brebis et la vache. Reprod. Nutr. Dévelop., 27, 327-398.

CHILLIARD Y., 1988. Rôles et mécanismes d'action de la somatotropine (hormone de croissance) chez le ruminant en lactation. Reprod. Nutr. Dévelop., 28, 39-59.

CHILLIARD Y., ROBELIN J., 1983. Mobilization of body proteins by early lactating cows measured by slaughter and D20 dilution techniques. IVth. Int. Symp. Protein metabolism and nutrition (Clermont-Ferrand), EAAP - Publ. N° 31, V°1. II, 195-198 (INRA Publ.).

COLLIER R.J., JOHNSON H.D., 1988. Bovine somatotropin - Mechanism of action and effects under differing environments. Monsanto technical Symposium, Fresno, California, USA, p 11-19.

DHIMAN T.R., KLEINMANS J., RADLOFF H.D., TESSMANN N.J., SATTER L.D., 1988. Effect of recombinant bovine somatotropin on feed intake, dry matter digestibility and blood constituents in lactating dairy cows. J. Dairy Sci., 71 (suppl.1), 121 (Abstract).

EISEMANN J.H., TYRRELL H.F., HAMMOND A.C., REYNOLDS P.J., BAUMAN D.E., HAALAND GL., McMURTRY J.P., VARGA G.A., 1986. Effect of bovine growth hormone administration on metabolism of growing hereford heifers: dietary digestibility, energy and nitrogen balance. J. Nutr., 116, 157-163.

EISENBEISZ W.A., CASPER D.P., SCHINGOETHE D.J., LUDENS F.C., 1988a. Lactational evaluation of recombinant bovine somatotropin with corn and barley diets : response to diets. J. Dairy Sci., 71 (suppl.1), 122 (Abstract).
EISENBEISZ W.A., CASPER D.P., SCHINGOETHE D.J., LUDENS F.C., SHAVER R.D., 1988b. Lactational evaluation of recombinant bovine somatotropin with corn and barley diets : response to somatotropin. J. Dairy Sci., 71 (suppl.1), 123 (Abstract).
ELVINGER F., HEAD H.H., WILCOX C.J., NATZKE R.P., EGGERT R.G., 1988. Effects of administration of bovine somatotropin on milk yield and composition. J. Dairy Sci., 71, 1515-1525.
EPPARD P.J., BAUMAN D.E., McCUTCHEON S.N., 1985. Effect of dose of bovine growth hormone on lactation of dairy cows. J. Dairy Sci., 68, 1109-1115.
EPPARD P.J., LANZA G.M., HUDSON S., COLE W.J., HINTZ R.L., WHITE T.C., RIBELIN W.E., HAMMOND B.G., BUSSEN S.C., LEAK R.K., METZGER L.E., 1988. Response of lactating dairy cows to multiple injections of sometribove, USAN (recombinant methionyl bovine somatotropin) in a prolonged release system. Part I. Production response. J. Dairy Sci., 71 (suppl.1), 184 (Abstract).
FARRIES E., PROFITTLICH C., 1987. The influence of applicated bovine somatotropin on some metabolic criteria in dairy cows. 38th Ann. Meeting Europ. Ass. Anim. Prod. (Lisbon, Portugal.) p432.
FAVERDIN P., HODEN A., COULON J.B., 1987. Recommandations alimentaires pour les vaches laitières. Bull. Tech. C.R.Z.V. Theix, I.N.R.A. 70, 133-152.
FURNISS S.J., STROUD A.J., BROWN A.C.G., SMITH G., 1988. Milk production, feed intakes and weight change of autumn calving, flat rate fed dairy cows given two weekly injections of recombinantly derived bovine somatotropin (BST). Proc. Brit. Soc. Anim. Prod., Winter Meeting, Paper N°.1.
GRAVERT H.O., PABST K., WOLLNY C., 1988. Quoted by Peel et al., 1988.
HARD D.L., COLE W.J., FRANSON S.E., SAMUELS W.A., BAUMAN D.E., ERB H.N., HUBER J.T., LAMB R.C., 1988. Effect of long term sometribove, USAN (recombinant methionyl bovine somatotropin), treatment in a prolonged release system on milk yield, animal health and reproductive performance-pooled across four sites. J. Dairy Sci., 71 (suppl.1), 210 (Abstract).
HART I.C., 1988. Altering the efficiency of milk production of dairy cows with somatotropin. In: Garnsworthy P.C. ed. Nutrition and lactation in the dairy cow. Butterworths, London, 232-247.
HEMKEN R.W., HARMON R.J., SILVIA W.J., HEERSCHE G., EGGERT R.G., 1988. Response of lactating dairy cows to a second year of recombinant bovine somatotropin (BST) when fed two energy concentrations. J. Dairy Sci., 71 (suppl.1), 122 (Abstract).

HUBER J.T., 1987. The production response of BST : feed additives, heat stress and injection intervals. In "National invitational workshop on bovine somatotropin". (St Louis, USA), 57-60.

HUBER J.T., WILLMAN S., MARCUS K., THEURER C.B., 1988. Effect of sometribove (SB). USAN (recombinant methionyl bovine somatotropin) injected in lactating cows at 14-d intervals on milk yields, milk composition and health. J. Dairy Sci., 71 (suppl.1), 207 (Abstract).

HUTCHISON C.F., TOMLINSON J.E., McGEE W.H., 1986. The effects of exogenous recombinant or pituitary extracted bovine growth hormone on performance of dairy cows. J. Dairy Sci., 69 (suppl.1), 152 (Abstract).

JENNY B.F., ELLERS J.E., TINGLE R.B., MOORE M., GRIMES L.W., ROCK D.W., 1988. Responses of dairy cows to recombinant bovine somatotropin in a sustained release vehicle. J. Dairy Sci., 71 (suppl.1), 209 (Abstract).

JOURNET M., CHILLIARD Y., 1985. Influence de l'alimentation sur la composition du lait. I. Taux butyreux : facteurs généraux. Bull.Tech. C.R.Z.V. Theix, I.N.R.A. 60, 13-23.

KIK N., COOK R.M., 1986. Effects of bovine somatotropin and IsoPlus on milk production. J. Dairy Sci., 69 (suppl.1), 158 (Abstract).

LAMB R.C., ANDERSON M.J., HENDERSON S.L., CALL J.W., CALLAN R.J., HARD D.L., KUNG L., 1988. Production response of Holstein cows to sometribove USAN (recombinant methionyl bovine somatotropin) in a prolonged release system for one lactation. J. Dairy Sci., 71 (suppl.1), 208 (Abstract).

LANZA G.M., BAILE C.A., COLLIER R.J., 1988a. Development and potential of BST. Nutr. Institute, Nat. Feeds. Ingr. Ass. (USA) (14p.).

LANZA G.M., EPPARD P.J., MILLER M.A., FRANSON S.E., GANGULI S., HINTZ R.L., HAMMOND B.G., BUSSEN S.C., LEAK R.K., METZGER L.E., 1988b. Response of lactating dairy cows to multiple injections of sometribove, USAN (recombinant methionyl bovine somatotropin) in a prolonged release system. Part III. Changes in circulating analytes. J. Dairy Sci., 71 (suppl.1), 184 (Abstract).

LANZA G.M., WHITE T.C., DYER S.E., HUDSON S., FRANSON S.E., HINTZ R.L., DUQUE J.A., BUSSEN S.C., LEAK R.K., METZGER L.E., 1988c. Response of lactating dairy cows to intramuscular or subcutaneous injection of sometribove, USAN (recombinant methionyl bovine somatotropin) in a 14-day prolonged release system. Part II. Changes in circulating analytes. J. Dairy Sci., 71 (suppl.1), 195 (Abstract).

LEITCH H.W., BURNSIDE E.B., MacLEOD G.K., McBRIDE B.W., KENNEDY B.W., WILTON J.W., 1987. Genetic and phenotypic affects of administration of recombinant bovine somatotrophin to holstein cows. J. Dairy Sci., 70 (suppl.1), 128 (Abstract).

LOSSOUARN J., 1988. Etude d'une formulation retard de zinc-methionyl bovine somatotropine pour la production laitière. Compte-rendu d'essai. INA Paris-Grignon-Monsanto-France.

LOUGH D.S., MULLER L.D., KENSINGER R.S., SWEENEY T.F., GRIEL Jr L.C., 1988. Effect of added dietary fat and bovine somatotropin on the performance and metabolism of lactating dairy cows. J. Dairy Sci., 71, 1161-1169.

LYNCH J.M., BARBANO D.M., BAUMAN D.E., HARTNELL G.F., 1988. Influence of sometribove (recombinant methionyl bovine somatotropin) on the protein and fatty acid composition of milk. J. Dairy Sci., 71 (suppl.1), 100 (Abstract).

MARSH W.E., GALLIGAN D.T., CHALUPA W., 1987. Making economic sense of bovine somatotropin use in individual dairy herds. In "Nutrient partitioning", Am. Cyanamid Techn. Symp., (California, USA), 59-79.

MEYER R.M., McGUFFEY R.K., BASSON R.P., RAKES A.H., HARRISON J.H., EMERY R.S., MULLER L.D., BLOCK E., 1988. The effect of somidobove sustained release injection on the lactation performance of dairy cattle. J. Dairy Sci., 71 (suppl.1), 207 (Abstract).

MOLLETT T.A., DeGEETER M.J., BELYEA R.L., YOUNGQUIST R.A., LANZA G.M., 1986. Biosynthetic or pituitary extracted bovine growth hormone induced galactopoiesis in dairy cows. J. Dairy Sci., 69 (suppl.1), 118 (Abstract).

MUNNEKE R.L., SOMMERFELDT J.L., LUDENS E.A., 1988. Lactational responses of dairy cows to recombinant bovine somatotropin. J. Dairy Sci., 71 (suppl.1), 206 (Abstract).

McDANIEL B.T., HAYES P.W., 1988. Absence of interaction of merit for milk with recombinant bovine somatotropin. J. Dairy Sci., 71 (suppl.1), 240 (Abstract).

McGUFFEY R.K., GREEN H.B., BASSON R.P., 1987b. Performance of Holsteins given bovine somatotropin in a sustained delivery vehicle. Effect of dose and frequency of administration. J. Dairy Sci., 70 (suppl.1), 177 (Abstract).

McGUFFEY R.K., GREEN H.B., BASSON R.P., 1988. Protein nutrition of the somatotropin-treated cow in early lactation. J. Dairy Sci., 71 (suppl.1), 120 (Abstract).

McGUFFEY R.K., GREEN H.B., FERGUSON T.H., 1987a. Lactation performance of dairy cows receiving recombinant bovine somatotropin by daily injection or in a sustained release vehicle. J. Dairy Sci., 70 (suppl.1), 176 (Abstract).

NYTES A.J., COMBS D.K., SHOOK G.E., 1988. Efficacy of recombinant bovine somatotropin injected at three dosage levels in lactating dairy cows of different genetic potentials. J. Dairy Sci., 71 (suppl.1), 123 (Abstract).

OLDENBROEK J.K., GARSSEN G.J. FORBES A.B., JONKER L.J., 1987. The effect of treatment of dairy cows of different breeds with recombinantly derived bovine somatotropin in a sustained delivery vehicle. 38th Ann. Meeting Europ. Ass. Anim. Prod., (Lisbon, Portugal). (32p).

PALMQUIST D.L., 1988. Response of high-producing cows given daily injections of recombinant bovine somatotropin from D 30-296 of lactation. J. Dairy Sci., 71 (suppl.1), 206 (Abstract).

PARRASSIN P.R., VIGNON B., 1988. Effects of somatotropin on milk production and composition in Montbéliarde cows - Unpublished results. INRA-Mirecourt-ENSAIA-Nancy-Elanco-France).

PEEL C.J., 1988. Bovine somatotropin (BST)- A review of efficacy and mechanism of action in dairy cows.. (To be published).

PEEL C.J., BAUMAN D.E., 1987. Somatotropin and lactation. J. Dairy Sci., 70, 474-486.

PEEL C.J., EPPARD P.J., HARD D.L., 1988. Evaluation of sometribove (methionyl bovine somatotropin) in toxicology and clinical trials in Europe and and the United States. (To be published).

PEEL C.J., FRONK T.J., BAUMAN D.E., GOREWIT R.C., 1982. Lactational response to exogenous growth hormone and abomasal infusion of a glucose-sodium caseinate mixture in high yielding dairy cows. J. Nutr., 112, 1770-1778.

PEEL C.J., SANDLES L.D., QUELCH K.J., HERINGTON A.C., 1985. The effects of long-term administration of bovine growth hormone on the lactational performance of identical-twin dairy cows. Anim. Prod., 41,135-142.

PELL A.N., TSANG D.S., HUYLER M.T., HOWLETT B.A., KUNKEL J., 1988. Responses of Jersey cows to treatment with sometribove, USAN (recombinant methionyl bovine somatotropin) in a prolonged release system. J. Dairy Sci., 71 (suppl.1), 206 (Abstract).

PHIPPS R.H., 1987. The use of prolonged release bovine somatotropin in milk production. Int. Dairy Fed. Congr. (Helsinki, Finland) (23p).

REMOND B., 1985. Influence de l'alimentation sur la composition du lait de vache-2-taux protéique: facteurs généraux. Bull. Tech. CRZV Theix, INRA, 62, 53-67.

REMOND B., CHILLIARD Y., CISSE M., 1988. Effects of somatotropin (sometribove) on feed intake, performances and metabolism of dairy cows fed two levels of concentrates . Unpublished results. (INRA-Theix- Monsanto-France).

RIJPKEMA Y.S., PEEL C.J., 1988. Quoted by Peel et al., 1988.
RIJPKEMA Y.S., REEUWIJK L.V., PEEL C.J., MOL E., 1987. Responses of dairy cows to long-term treatment with somatotropin. 38th Ann. Meeting Europ. Ass. Anim. Prod. (Lisbon, Portugal). p428.
ROWE-BECHTEL C.L., MULLER L.D., DEAVER D.R., GRIEL L.C., 1988. Administration of recombinant bovine somatotropin (rbSt) to lactating dairy cows beginning at 35 and 70 days postpartum. I. Production response. J. Dairy Sci., 71 (suppl.1), 166 (Abstract).
SAMUELS W.A., HARD D.L., HINTZ R.L., OLSSON P.K., COLE W.J., HARTNELL G.F., 1988. Long term evaluation of sometribove, USAN (recombinant methionyl bovine somatotropin) treatment in a prolonged release system for lactating cows. J. Dairy Sci., 71 (suppl.1), 209 (Abstract).
SCHNEIDER P.L., VECCHIARELLI B., CHALUPA W., 1987. Bovine somatotropin and ruminally inert fat in early lactation. J. Dairy Sci., 70 (suppl.1), 177 (Abstract).
SECHEN S.J., DUNSHEA F.R., BAUMAN D.E., 1988. Mechanism of bovine somatotropin (bST) in lactating cows:effect on response to homeostatic signals (epinephrine and insulin). J. Dairy Sci., 71 (suppl.1), 168 (Abstract).
SODERHOLM C.G., OTTERBY D.E., LINN J.G., EHLE F.R., WHEATON J.E., HANSEN W.P., ANNEXSTAD R.J., 1988. Effects of recombinant bovine somatotropin on milk production, body composition, and physiological parameters. J. Dairy Sci., 71, 355-365.
SULLIVAN J.T., TAYLOR R.B., HUBER J.T., FRANSON S.E., HOFFMAN R.G., HARD D.L., 1988. Relationship of production level and days postpartum to response of cows to sometribove, USAN (recombinant methionyl bovine somatotropin). J. Dairy Sci., 71 (suppl.1), 207 (Abstract).
TESSMANN N.J., KLEINMANS J., DHIMAN T.R., RADLOFF H.D., SATTER L.D., 1988. Effect of dietary forage:grain ratio on response of lactating dairy cows to recombinant bovine somatotropin. J. Dairy Sci., 71 (suppl.1), 121 (Abstract).
THOMAS C., JOHNSSON I.D., FISHER W.J., BLOOMFIELD G.A., MORANT S.V., WILKINSON J.M., 1987. Effect of somatotrophin on milk production, reproduction and health of dairy cows. J. Dairy Sci., 70 (suppl.1), 175 (Abstract).
TYRRELL H.F., BROWN A.C., REYNOLDS P.J., HAALAND G.L., PEEL C.J., BAUMAN D.E., STEINHOUR W.C., 1982. Effect of growth hormone on utilization of energy by lactating Holstein cows. In EKERN A., SUNDSTOL F.,"Energy metabolism of farm animals". EAAP. Publ. N° 29, 46-47.
VEDEAU F., SCHOCKMEL L.R., 1988. Quoted by Peel et al., 1988.

VERITE R., RULQUIN H., FAVERDIN P., 1988. Effects of slow released somatotropin on dairy cow performances (see this Seminar).

VERNON R.G., 1986. The response of tissues to hormones and the partition of nutrients during lactation. Hannah Res. Inst., Rep. 1985, 115-121.

VIGNON B., LAURENT J.F., WALLET P., 1988. Effects of somatotropin on milk production and composition in cows at pasture - Unpublished results. (ENSAIA-INRA-Nancy-Elanco-France).

WEST J.W., JOHNSON J.C.Jr., BONDARI K., 1988. The effect of bovine somatotropin on productivity and physiologic responses of lactating Holstein and Jersey cows. J. Dairy Sci., 71 (suppl.1), 209 (Abstract).

WHITAKER D.A., SMITH E.J., KELLY J.M., HODGSON-JONES L.S., 1988. Health, welfare and fertility implications of the use of bovine somatotrophin in dairy cattle. Vet. Rec., 122, 503-505.

WHITE T.C., LANZA G.M., DYER S.E., HUDSON S., FRANSON S.E., HINTZ R.L., DUQUE J.A., BUSSEN S.C., LEAK R.K., METZGER L.E., 1988. Response of lactating dairy cows to intramuscular or subcutaneous injection of sometribove, USAN (recombinant methionyl bovine somatotropin) in a 14-day prolonged release system. Part I. Animal Performance and Health. J. Dairy Sci., 71 (suppl.1), 167 (Abstract).

A REVIEW OF THE INFLUENCE OF SOMATOTROPIN ON HEALTH, REPRODUCTION AND WELFARE IN LACTATING DAIRY COWS

R.H. Phipps

AFRC Institute for Grassland and Animal Production
Church Lane, Shinfield, Reading, RG2 9AQ, U.K.

ABSTRACT

The administration of BST to lactating dairy cows did not markedly affect clinical measurements such as body temperature or respiration rate but heart rate was increased slightly at high dose rates. Although some trials have recorded an increase in leg disorders, the majority of experiments have not experienced this problem. There does not appear to be a depletion of calcium or phosphorus from bones of treated animals. BST produced no consistent effect on either the incidence of clinical mastitis or milk somatic cell counts (SCC); while SCC were generally higher for treated animals, few of the differences reached significance. BST may positively influence the restoration of mammary function following mastitis. Blood chemistry changes that occurred with treatment were generally related to increased production levels and remained within accepted clinically normal ranges. The decrease in haemoglobin and haematocrit came in this category, and were too small to reflect functional anaemia. There was no indication of either subclinical or clinical ketosis or "burn-out". Calf health, growth and subsequent development were unaffected by treatment.

It would appear that high dose rates of BST used in early lactation are more likely to decrease reproductive efficiency than lower dose rates used later in lactation. The use of currently suggested dose rates administered from day 80 of lactation did not appear to have any substantive effect on reproductive efficiency. Where days open were increased it was clearly established as a level of production effect and not attributed to BST per se. Any potential effects on reproduction can be minimised by delaying the start of treatment until a high proportion of cows had conceived. This could well happen as BST might be used for short term strategic use such as quota control or seasonal production.

The considered opinion of veterinary surgeons was that there was no evidence of any welfare problems arising from the administration of BST. In general, cows did not show reluctance to enter the handling facilities where injections were carried out. The subcutaneous injection caused a mild reaction which with one exception was considered insensitive and were resolved in 2-4 weeks. If BST leads to fewer cows/herd it might actually lead to improved animal welfare with more available stockman time/cow. It seems extremely unlikely that "stressed" cows would respond in the manner of BST treated cows. Chronic and acute overdosing did not appear to lead to problems. The suggestion of the Farm Animal Welfare Council of producing a set of guidelines for commercial administration of BST including monitoring of herd health and performance and that this should be done in conjunction with the Royal College of Veterinary Surgeons merits consideration.

HEALTH

Physical Examinations

Clinical measurements show that body temperature and respiration rate are unaffected by the administration of BST (Soderholm et al., 1988; Phipps & Weller, 1988; Eppard et al., 1987), but heart rate tended to be raised slightly at high dose rates (Soderholm et al., 1988) (Table 1).

Physical examinations have revealed various minor abnormalities in cows in both control and treated groups (Phipps et al., 1988). The most common was fluid filled bursal swellings of the carpus and tarsus, which became more frequent in both groups as lactation progressed with a higher incidence in the treated group (Table 2). Such swellings are not uncommon in housed cattle and are usually resolved at pasture, which could not happen in this case as cows were housed throughtout the year. In the same study the incidence of lameness in the first lactation was similar for control and treated cows but was higher in treated cows in the second lactation. When considering these data it should be remembered that these animals had been housed for approximaely 2 years.

In a commercial herd in the U.S.A., however, BST did not increase the incidence of leg disorders (Chalupa et al., 1988). Farm trials have also been carried out in the U.K. and show that BST does not necessarily lead to an increased incidence of lameness, which may be more closely associated with individual farms and prevailing management practices (Table 3).

In order to examine the long term effects of BST on bone growth and development, 12 cows in their third lactation were examined (Phipps, unpublished data). The cows chosen had received either BST or placebo for two consecutive lactations and had all been 2 year old calving heifers at the start of treatment. Their left and right carpi were radiographed. The interpretation of the radiographs made by S. Dean, (MRCVS, Dip. Vet. Radiology) revealed normal bone growth, development, density and periosteal activity in BST treated cows with no treatment related abnormalities. These data indicate that bone development is normal and that metabolic balance is maintained in BST treated cows and does not suggest that bones become depleted of calcium and phosphorus.

TABLE 1   Clinical measurements during a 2 year efficacy study

	Body temp (°C)	Respiratory (rate/min)	Heart (rate/min)
Lactation 1			
Pretreatment			
Control	38.8	36	52
BST	38.8	36	55
Experimental			
Control	38.8	29	82
BST	38.8	30	84
Lactation 2			
Pretreatment			
Control	38.9	30	80
BST	38.8	31	88
Experimental			
Control	38.8	29	95
BST	38.9	34	91

(Phipps & Weller, 1988)

TABLE 2   Incidence of swollen carpus and lameness

	Swollen carpus/tarsus	Lame/Abnormal gait
Lactation 1 (n = 45/group)		
Pretreatment		
Control	3	2
BST	2	5
Experimental Period		
Control	7	4
BST	6	5
Dry Period		
Control	6	3
BST	10	2
Lactation 2 (n = 30/group)		
Pretreatment		
Control	3	5
BST	4	2
Experimental Period		
Control	4	4
BST	7	7
Dry Period		
Control	7	7
BST	10	10

(Phipps & Weller 1988)

TABLE 3 Incidence of lameness in the U.K. commercial trials

Trial	No. of cows lame
1 Control (n = 30)	7
BST (n = 60)	14
2 Control (n = 28)	9
BST (n = 44)	3
3 Control (n = 33)	0
BST (n = 60)	2

n = nos. of cows/group   (Monsanto plc, Personal Communication)

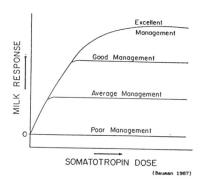

Figure 1   Milk Response to Somatotropin Dose

Metabolic Disorders

Kronfeld (1987) and others predicted that the use of BST would have a catastrophic effect on animal health leading to problems of ketosis, milk fever and "burn out" of treated cows. This would appear not to be the case. Bauman (1987) reported that "since the first study in 1937 over a hundred refereed papers had been published relating to the use of BST, and not one of these papers referred to a ketosis or milk fever problem". Since Bauman's article at least a further 50 papers have appeared in press and these too have reported that the administration of BST did not predispose animals to either ketosis or milk fever. Ketosis was not even a problem when treated cows received pasture alone (Peel et al., 1985).

Indeed our own current multi-lactational trial with cows now in their fourth consecutive lactation of BST treatment has not produced a single case of ketosis (Phipps & Weller 1988).

The administration of BST does not even increase the incidence of subclinical ketosis as measured by B-hydroxybutyrate concentration in the blood, (Table 4) (Hutchison et al., 1986; Eppard et al., 1987; Lanza et al., 1988; Dhiman et al., 1988; Phipps, 1988).

To understand why cows can produced 20%+ more milk when treated with somatotropin without any increase in metabolic diseases the review by Peel & Bauman (1988) should be read in detail. It seems that somatotropin exerts a homeorhetic control and co-ordinates a whole range of physiological mechanisms. Treated cows appear to bear a strong similarity to "genetically superior cows", in which for example feed intake is adjusted to meet increased milk yield. Thus somatotropin does not override physiological mechanisms and make a medium yielding cow produce more milk, in which circumstances an increase in metabolic problems might be expected. Rather it allows the "genetically superior cow" to exhibit its potential.

There is evidence to suggest that the response obtained from these "genetically superior cows" is influenced by the prevailing level of management. To quote Asimov & Krouze (1937) ... "is more effective in well-kept cattle." Bauman (1987) has also postulated that the response to BST will depend on management conditions, and that there will be a diminishing response curve at each management level (Fig. 1). The work of Mollett et al., (1986) supports this hypothesis, as when environment, nutrition and management were all below optimum, the response to BST was also low. What is of paramount importance, however, is that there were no adverse effects on the cows. Another example which shows that BST does not exert an overriding influence on milk yield is that the response to BST decreases as the natural end of lactation approaches (Table 5).

Table 4 Concentration of B-hydroxybutyrate in blood of control and treated cows

Sample	B-hydroxybutyrate (mmol/l)	
	Control	BST
1	0.68	0.61
2	0.58	0.45
3	0.71	0.58
4	0.75	0.63
5	0.58	0.59
Average	0.66	0.57

Normal range <0.7

Table 5 Response to BST as lactation progresses

Week of Lactation	Milk Yield (kg/d)
9 - 17	+4.5
18 - 25	+4.4
26 - 33	+3.6
34 - 41	+2.5

(Phipps unpublished data)

Blood Chemistry

Blood profiles of control and treated cows receiving 0, 13.5, 27.0 and 40.5 mg/d of BST have been determined (Eppard et al., 1987). The averaged results for somatotropin treatments showed that values for treated and control cows were in general similar, and within the normal range. However, the authors reported that, particularly at the highest dose rate haematocrit (%) and haemoglobin (g/dl) were decreased compared with the controls. A similar finding has been reported for haematocrit (%) by Soderholm et al. (1986). However, work by Dhiman et al., (1988) showed that when using a dose rate, more in keeping with practical application (20.6mg/d) blood measurements were unaffected.

Analyses have been carried out on cows receiving a prolonged release formulation (500 mg/14d) which showed a similar pattern to that recorded with daily injections. Packed cell volume, haemoglobin and red blood cell values were all lower for treated when compared with control cows (Table 6), although all values were within the normal range.

Unless the decreases in the indices are very large it is not possible to infer that these changes represent true anaemia (Allen, Personal Communication 1988). The changes in these three haematological indices may only reflect a shift in the balance between the total value of red cells and total volume of plasma. These in turn are known to be influenced by factors such as yield and plane of nutrition and the changes observed are in the direction expected for higher yielding cows.

Other minor changes in blood metabolites have been recorded but these have reflected metabolic shifts to support increased milk production. (Cole et al., 1988; Lanza et al., 1988)

TABLE 6 Blood Analyses

	Packed Cell Volume (5)	Haemoglobin (g/dl)	Red Blood Cells ($10^{12}/1$)
Sample 1 Control (n = 6)	31.0	11.1	5.78
BST (n = 6)	28.1	10.3	5.60
2 Control	31.2	11.6	5.90
BST	26.9	10.0	5.43
3 Control	30.9	12.0	6.12
BST	26.0	9.8	5.51
4 Control	29.7	11.9	6.17
BST	24.8	10.0	5.51
5 Control	27.5	10.9	5.55
BST	24.8	9.7	5.36
Average Control	30.1	11.5	5.90
BST	26.1	10.0	5.48
Normal Range	24-40	8-14	5-9

(Phipps unpublished data)

Udder Health
(a) Somatic Cell Counts (SCC)

Although numerous factors influence SCC it is generally accepted as an indicator of udder health and the level of subclinical mastitis.

In a 2-year study carried out in the U.K. the SCC for both control and BST treated cows were low (Phipps & Weller, 1988), being less than half the national average (Booth 1988). However, in both years the values recorded for the treated groups were significantly higher than the controls. A similar but non-significant trend of increased SCC in treated cows has also been reported from Holland (Rijpkema et al., 1987).

However, recent trials in North America reported that SCC for treated cows were similar to that of control cows (McGuffey et al., 1987; Elvinger et al., 1987; Hemken et al., 1988; Rowe-Bechtel et al., 1988; Munneke et al., 1988; Hard et al., 1988; Eppard et al., 1987; Cole et al., 1988; Aguilar et al., 1988).

Nevertheless in a recent review by Peel et al., (1988) the authors reviewed the SCC from eight clinical trials in the U.S.A. and EEC (Table 7). Treated cows (500 mg/14d) tended to have higher SCC compared with controls, but the differences were only significant at two of the eight sites (U.K. and New York).

(b) Clinical mastitis

Mastitis is a major production disease. The cost to the dairy industry is high through lost production and enforced culling (Booth, 1988).

Work in the U.S.A. at Cornell (Eppard et al., 1987) showed no increased incidence of mastitis even when BST was used at the highest dose rate of 40.5mg/d. More recent papers although only published in abstract form at the 1988 American Dairy Science Association meeting support this finding (Rowe-Bechtel et al., 1988; Palmquist, 1988; Lamb etc. al., 1988; Hard et al, 1988; Chalupa et al., 1988). However, an even more recent review by Peel et al., (1988) provides details of the incidence of clinical mastitis in efficacy studies carried out in the U.S.A. and EEC. (Table 8). In Table 8 information has also been included from the second year of study reported by Phipps & Weller (1988) and also from a range of additional studies carried out on experimental units and commercial herds in the U.K.

From the available data the differences between control and treated cows are not clear cut. In certain circumstances there appears to be an increased incidence of clinical mastitis in treated cows yet in other cases there is no indication of increased clinical mastitis as a result of BST treatment. However, the overall incidence of clinical mastitis was notably also higher before BST treatment commenced in cows already allocated to the treatment group and thus the relative incidence of mastitis was not affected by BST treatment (Table 8).

Professor Burvenich has kindly given permission for some of his, as yet unpublished, work on mastitis and BST to be referred to. After experimentally infecting half the udder with E. coli, treatment of either antibiotic therapy alone or in conjunction with BST was implemented. While there was no decrease in lactose concentration in uninfected quarters milk yield declined to a similar extent in both treatments, but only returned to pre-infection levels in quarters treated with antibiotic and BST. In the infected quarters the decline in milk yield was less and recovery as determined by the return to a normal lactose concentration was quicker for the quarters treated with antibiotic and BST compared with antibiotic therapy alone. Under the experimental conditions of E. coli mastitis short term treatment with BST appears to have beneficial effects upon the recovery from the disease. Further studies involving longer term BST treatments would be of interest.

TABLE 7 Mean somatic cell counts

	Somatic Cells (x 10^3/ml)			
	Pretreatment		Treatment	
	0	500*	0	500
Arizona	167	184	193	250
New York	55	96	97	221
Missouri	179	144	202	250
Utah	150	65	284	191
France	153	172	153	178
Germany	140	124	192	264
Netherlands	107	148	137	186
United Kingdom	60	60	133	196

(Peel et al., 1988) *500mg/d BST

TABLE 8 Incidence of clinical mastitis

	Total Cows		Number of cows with clinical mastitis			
			Pretreatment		Treatment	
Clinical Trials	Control	BST*	Control	BST	Control	BST
Arizona	40	40	2	4	10	14
New York	40	40	0	1	4	14
Missouri	63	63	18	20	17	19
Utah	36	36	2	3	11	7
France	29	29	4	10	5	8
Germany	30	30	5	6	3	2
Netherlands	32	32	1	0	5	7
U.K.						
Year 1	45	45	3	4	3	3
Year 2	30	30	2	4	1	8
Total affected			37	52	59	82
BST : Control			1.41		1.39	
Experimental Units in the U.K.						
Bridgets EHF	12	12	0	0	2	2
West of Scotland College of Agriculture	16	16	-	-	0	2
Edinburgh University	18	20	-	-	3	6
Commercial Herds in U.K.						
1	33	33	4	1	1	1
2	45	45	8	7	8	11
3	55	54	7	5	3	6
4	30	60	-	-	2	1
5	28	44	-	-	1	1
6	33	60	-	-	0	2

*500mg/14d

(Peel, et a., 1988; Phipps & Weller, 1988; Monsanto plc unpublished data)

Calf Health and Subsequent Growth and Skeletal Development

As soon as it was practically possible all calves were subjected to a thorough physical examination by a veterinary surgeon in the study reported by Phipps & Weller (1988). Body temperature, heart and respiration rates taken at these examinations were similar for calves from both control and BST treated cows (Table 9).

Other measurements taken on female calves kept as replacements were bodyweight, height at withers, heart girth, crown-rump length and length of metacarpus. These measurements recorded at birth and monitored during the subsequent eight weeks showed that calf growth and development was unaffected by the administrtion of BST to the dams (Table 9). Thomas et al. (1987) have also reported that calf birth weights were unaffected by treatment as has Eppard et al., (1987).

A number of the heifers kept as herd replacements were monitored during the subsequent two years. No differences were noted between these heifers in average liveweight gain, days to first service and serves/+PD (Table 10). (Phipps, unpublished data). These heifers will shortly enter the main herd where their performance will continue to be monitored.

TABLE 9 Measurements taken for calves at birth and eight weeks of age

	Lactation 1		Lactation 2	
	Control	Treated	Control	Treated
Clinical Measurements				
Rectal temperature (°C)	38.9	38.8	39.1	39.1
Pulse (beats/min)	148	139	143	145
Respiration rate (resps/min)	54	52	52	53
Measurements at birth				
Weight (kg)	36.3	39.9	37.7	37.5
Height at withers (cm)	74	75	75	76
Heart girth (cm)	76	78	76	77
Crown-rump length (cm)	90	93	90	93
Length of metacarpus (cm)	16	16	16	16
Average weekly growth				
(over first 8 weeks)				
Weight (kg)	4.2	3.8	3.5	3.7
Height at withers (cm)	1.2	1.1	0.7	1.1
Heart girth (cm)	2.5	1.9	1.8	1.8
Crown-rump length (cm)	2.5	2.2	1.5	1.7
Length of metacarpus (cm)	0.2	0.1	0.2	0.2

TABLE 10 Performance of dairy herd replacements from control and BST treated cows

	Control	BST
Average L. wt gain (kg/d)	0.76	0.77
Days to 1st service	428	423
No. serves for + P.D.	1.92	1.88

REPRODUCTION

Reproductive performance in a dairy herd is important as a decline in efficiency can lead to reduced profit (Esslemont et al., 1985). As such the possible effects of BST must be examined in detail.

Production responses to BST have shown that whereas milk yield responds almost immediately to treatment, there is generally a time lag of 4-5 weeks before intake increases to meet the increased level of milk production (Phipps, 1988, Thomas et al., 1988). This leads to a marked change in energy balance status and often an increased energy deficit (Phipps, 1988, see Fig. 2). Ducker et al. (1985) have reported that such changes are likely to adversely affect reproductive efficiency. On this basis it might well be surmised that if BST was introduced in early lactation; at a time which coincided with the main rebreeding period, reproductive efficiency might be reduced. Thus time of introduction, dose rate and plane of nutrition, all of which would effect energy status, are likely to influence reproductive performance of treated cows. When considering the effects reported in the literature, it is important to remember that few if any of the trials contain sufficient animals for significant differences to be established. Thus underlying trends must be distinguished and their significance, if any, commented on.

Influence of dose rate and time of administration

Table 11 shows the effects of a range of BST dose rates (0, 12.5, 25 and 40-50 mg/d) when administration started at day 28 (Burton et al., 1988; Chalupa et al., 1988; Thomas et al., 1987). These trials indicated that at higher dose rates, days open increased and pregnancy rate decreased. However, days to first service and services/conception were apparently unaffected. However, when lower dose rates (0, 10.3, 20.6 or 30.9 mg/d) were examined on a commercial farm days open and pregnancy rate were not affected at the 10.3 and 20.6 mg/d dose rate, but were adversely affected at the highest rate (Chalupa et al., 1988). Somewhat conflicting evidence has, however, been reported where days open were increased even at 6.25, 12.0 and

Fig 2: **Effect of BST on milk yield, dry matter intake and energy balance**

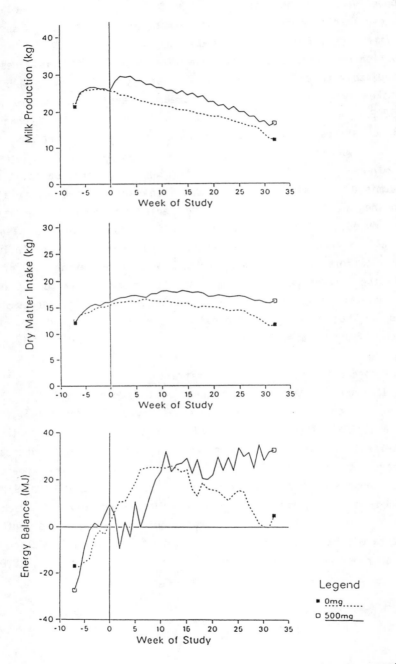

(Phipps 1988)

25.0 mg/d dose rates (Elvinger et al., 1988). These results may perhaps be attributed to the fact that there were only 10 cows/treatment, the response in terms of increased milk yield was high and the high ambient temperatures prevailing in Florida might have been a contributing factor to decreased reproductive efficiency. In the work reported by Eppard et al. (1987) the administration of BST did not start until day 84 of lactation. The results showed that at this later date of administration, days open and services/conception of cows treated with dose rates ranging from 13.5 - 40.5 mg/d were at least as good as the controls and comparable with the main herd (Table 11).

More emphasis is now being placed on the use of prolonged release formulations of BST rather than daily injections. Table 12 shows some of the reproductive measures recorded in six long term studies when BST was administered from day 60 of lactation at a dose rate of 500 mg/14d. When averaged over the six trials, days open were increased from 93 to 105 and services/conception from 1.88 to 2.12 for the treated cows when compared with the controls. However, Bauman et al. (1988) using the same product and administration date reported no effect on reproduction.

In three commercial herds in the U.K. where the introduction of BST (500 mg/14d) was instigated at approximately 80 days post partum, reproduction was essentially the same for both control and treated cows (Table 13).

Influence of Plane of Nutrition

Tessman et al., (1988) have reported an extremely elegant trial in which they examined the effect of concentrate (%) in the ration on reproductive performance of BST treated (20.6 mg/d) cows. When the ration contained 62, 52 and 32% concentrate in early, mid and late lactation services/conception and days open were similar for control and treated cows (Table 14). A decrease in concentrate input to 42, 32 and 12% of the ration in early, mid and late lactation influenced both control and treated groups, increasing services/conception and days open when compared with the higher concentrate input. However, the increase in services/conception and days open was greater for treated compared with the control cows (Table 14). This is probably a reflection of their lower energy status during the breeding period.

Days open - what causes the increase

It is generally well accepted that high yielding cows are more difficult to get in calf (Bourchier et al., 1987). An analysis of some 6.4 million lactations showed that an increase of 1 day open was associated with 100kg increase in 305-day lactation yield. (Huth & Schutzlar, 1987). The decrease in reproductive efficiency has been associated with an increased energy deficit in early lactation (Ducker et al., 1985). Thus a decrease in reproductive efficiency of cows treated with BST may be expected but is likely to be a consequence of their increased milk energy output rather than a direct effect of BST.

An analysis has been carried out to substantiate this suggestion. Data from four US and two EEC clinical trials were pooled. Cows were divided into quartiles based on 3.5% fat corrected milk during the first 252 days of lactation and then related to days open (Fig 3). The result showed that the increase in days open was significantly related to milk production and not to treatment. (Hard et. al., 1988)

Twins

There is an indication that the use of BST may increase the incidence of twinning above the 3% normally recorded. If a direct link is established then it adds support to the suggestion that the administration of BST should start after the major proportion of the cows had conceived, as twins are undesirable.

Fig. 3. Effect of milk production level on days open for control and treated cows.

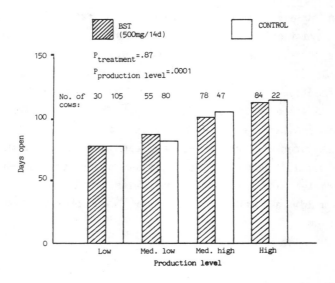

TABLE 11 The effect of BST dose rate and time of administration on reproductive performance

	\multicolumn{11}{c}{BST Dose Rate (mg/d)}										
Administered day 28	0	6.25	10.3	12.5	13.5	20.6	25	27	30.9	40.5	50
Burton et al., (1988)											
Days to serve	79			74			81				79
Days open	88			156			157				118
Chalupa et al., (1987)											
Services/conception	2.3			1.6			2.3				1.7
Preg. rate %	82			76			62				41
Elvinger et al., (1988)											
Days open	116	142		142			143				
Chalupa et al., (1988)											
Days open	122		116			113			131		
Preg. rate (%)	87		90			83			77		
Administered day 84											
Eppard et al., (1987)											
Days open	146				124			126		117	
Service/conception	3.2				2.0			2.4		2.0	

TABLE 12 Effect of prolonged release formulation of BST (500 mg/14d) on reproductive performance

	Control	BST
Furniss et al., (1988)		
Days open	115	113
Services/conception	2.3	1.8
Pell et al., (1988)		
Days open	79	91
Services/conception	1.6	2.2
Preg. rate (%)	96	86
Huber et al., (1988)		
Days open	90	105
Preg. rate (%)	81	75
Phipps & Weller (1988)		
Year 1 Days open	93	113
Services/conception	2.0	2.5
Preg. rate (%)	93	91
Year 2 Days open	81	88
Services/conception	1.8	2.3
Preg. rate (%)	90	83
Rijpkema et al., (1988)		
Days open	100	119
Services/conception	1.7	1.8

TABLE 13 Effect of prolonged release formulation of BST (500 mg/14d) on reproductive performance in commercial herds in the U.K.

	Treatment days post calving	% Pregnant		Services/conception		Days open	
		Control	BST	Control	BST	Control	BST
Herd 1	66 - 86	96 (n = 27)	93 (n = 27)	1.73	1.56	95	91
2	80 ± 7	92 (n = 38)	83 (n = 41)	1.31	1.38	97	103
3	80 ± 7	85 (n = 54)	81 (n = 52)	1.76	1.38	82	78
Average		91 (n = 119)	86 (n = 120)	1.60	1.44	91	91

(Monsanto plc 1988; Personal Communication)

TABLE 14 Effect of plane of nutrition on the reproductive performance of BST (20.6 mg/d) treated cows

Stage of Lactation	% Concentrate in ration	Services/conception		Days Open	
		Control	BST	Control	BST
Week 1 - 12	62				
13 - 26	52	1.7	1.8	94	95
27 - 44	32				
Week 1 - 12	42				
13 - 26	32	2.6	3.2	99	120
27 - 44	12				

(Tessman et al., 1988)

WELFARE

Response of the Farm Animal Welfare Council in the U.K.

The Farm Animal Welfare Council in the U.K. was asked to consider the use of BST. Their reply to the Minister of Agriculture was published in Hansard on 17th December, 1987. The reply stated that, "there is at present no evidence of any welfare problems arising from the use of BST in the short term. There are, however, areas of uncertainty which could have welfare implications where additional scientific information may be required in order to take a more considered opinion." These areas are shown in Table 15 together with references to recent work which goes a long way to answering the questions.

In addition the reply from the Farm Animal Welfare Council went on, "there appears to be a number of potential welfare abuses which could arise." These are shown in Table 16, with comments that would support the Councils recommendation that "guidelines for the commercial administration of BST, including monitoring of herd performance and health should be drawn up in conjunction with the Royal College of Veterinary Surgeons and that its use should be subject to veterinary supervision.

TABLE 15 Areas of uncertainty raised by the Farm Animal Welfare Council in the U.K. and the reference/response to appropriate work

Area	Reference/Response
(a) time of administration to high yielding cows	Bauman et al., (1984), Rowe-Bechtel et al. (1988)
(b) effect of BST on long term reproductive efficiency	The current paper examines this point
(c) response of cows differing in body composition and on different levels of energy and protein.	Tessman et al., (1988). McGuffey et al., (1988)
(d) potential skeletal problems in heifers	Phipps & Weller (1988) - None
(e) performance of calves born from treated animals	Phipps & Weller (1988) - Normal
(f) effect that exogenous BST may have on the ability of the animal to produce endogenous BST and thus on subsequent unsupplemented performance	Stop/start trials suggest no problems. Response of pituitary to GHRF is unchanged in BST supplemented cows compared with controls (Wollny, Personal Communication).
(g) use of BST over a number of lactations in the same animals	Numerous 2-lactation studies have now been reported with no problems. The study at I.G.A.P. is entering its fourth consecutive lactation of BST treatment of the same animals.

TABLE 16 Farm Animal Welfare Councils areas of concern over a number of potential welfare abuses

	Area	Response
(a)	use of BST to exploit overstretched cull cows by extending lactation	It is difficult to imagine the barren/cull cow as being overstretched at the end of lactation
(b)	lack of appropriate facilities for injection on some farms	This should be relevant to any veterinary procedure carried out on the farm
(c)	administration of BST without adequate veterinary supervision	Adequate veterinary supervision should be available
(d)	administration of BST to cows whose milk yield was depressed by reasons of undiagnosed disease or metabolic disorders	Burvenich has shown for mastitis that milk yield is not increased in the uninfected quarters when BST is administered. The response in the case of metabolic disorders is unknown
(e)	inadequacy of nutritional levels in some systems	Evidence suggests that under these circumstances the response to BST would be small. However, further data in this area would be desirable. Physiological studies support the above observation, showing that target tissues of underfed ruminants are less responsive to BST (Gluckman et al., 1987).

Needs and requirements of farm animals

These are outlined in the U.K. by the Farm Animal Welfare Council and are:
(a) Proper nutrition
(b) Thermal and physical comfort
(c) Freedom from pain, injury or disease
(d) Freedom from fear or distress
(e) Freedom to express most normal, socially acceptable behaviour patterns

These five points will form the basis for further discussion on the effect of BST on animal welfare.

(a) Proper nutrition

In order to obtain the expected response from BST, cows should be managed as though they were potential high yielders, and their diets formulated accordingly. However, if for some reason the plane of nutrition is inadequate, there is evidence that the response to BST will be less, but there is no suggestion of an increase in disease related incidences. Nevertheless further evidence to support Bauman's (1987) suggested relationship between level of management, response to BST and health related incidences would be desirable. These data may already exist from commercial herd trials and if it does it should be published as soon as possible. Although body condition scores have been shown to be lower during treatment with BST (Phipps, 1988) it must be remembered that it was recovered during the dry period, so that prior to the subsequent experimental period treated and control cows were in similar body condition. Once again changes in body condition score in adverse nutritional conditions should be explored.

(b) Thermal and physical comfort

This refers to the buildings and environmental conditions existing at the time of treatment. Experiments in climatic chambers have shown that good responses to BST have been recorded at both low and high temperatures, with no change in rectal temperature (Peel, 1988). This indicates that BST may have applications in countries outside the temperature zone.

With both clinical trials and those carried out in commercial herds, veterinary surgeons were asked for and gave their opinions on welfare issues relating to BST. Their comments are particularly relevant to sections (c) and (d) in the code of the Farm Animal Welfare Council.

(c) Freedom from pain, injury or disease

Does the injection of BST cause undue pain?

The response of the veterinary surgeons has been quite clear in that they considered the injection to be no more painful than any other injection. The quantity of material injected is small (1.4ml/14d). There is an indication that certain sites may be more sensitive than others. If this is so then clear guidelines for injection sites should be drawn up. This perhaps could be done in conjunction with the Royal College of Veterinary Surgeons.

The use of BST as an injection for non-therapeutic purposes does not set a precedent e.g. Cows with low reproductive performance also receive non-therapeutic injections to overcome this production problem for farmers.

Does the injection of BST cause injury?

The injection does cause a mild reaction in most animals which is generally resolved in 2-4 weeks. It is perhaps opportune to clarify an apparent discrepancy in the published literature. In the Veterinary Record, Phipps et al., (1988) reported that transient swellings were resolved in 2-4 days, while Whitaker et al., (1988) reported that swellings took 2-4 weeks to resolve. This difference was due to that fact that Phipps et al., (1988) were reporting on the first year of a clinical trial in which the intramuscular route for injection was used compared with a subcutaneous route by Whitaker et al., (1988). Of the 10 independent veterinary opinions that I have had access to only one (Whitaker et al., 1988) considered that the swellings were slightly tender to the touch. None of these swellings were, however, noted by the stockmen. The actual injection of BST did not lead to abscesses or complications.

Does the injection of BST cause disease?

As Webster (1985) said good health must assume first place in the hierarchy of welfare needs. "It would be a retrograde step to introduce a system that allowed animals perhaps more freedom of behaviour if for example it markedly increased the incidence of disease or untimely death." The results of trials reviewed in earlier sections showed that treated cows were generally in excellent health throughout experiments and that BST could not be linked to a substantive increase in disease, and the absence of metabolic diseases such as ketosis was notable. It would, nevertheless be desirable to have further information when BST was used in adverse management conditions. Conversely, the work of Professor Burvenich, reported earlier in this paper, which indicted that BST helped animals to recover more quickly from mastitis is of great interest and clearly further results are eagerly awaited.

(d) Freedom from fear and distress

Of the 10 veterinary opinions that I have read two stated that cows showed reluctance to enter handling facilities for future injections. The veterinary surgeons considered that animals were anxious about restraint and handling rather than what was injected.

In our own trial now entering its fourth consecutive lactation of treatment, cows have never shown any reluctance to enter the handling facilities in which the injections were given. I think that it is important for farmers to have adequate handling facilities not just for the possible use of BST but for the many and varied needs of the dairy herd.

(e) Freedom to express most normal, socially acceptable behaviour patterns

There is no indication that the use of BST affects behavioural patterns. Clearly if a farmer were to use BST for a substantial part of the lactation, herd size might decrease. In these circumstances the cows welfare may be improved as "stockmanship"/cow should increase.

A more stable social hierarchy within the herd may also be favoured by a reduction in group size.

Other related animal welfare issues

The misuse and abuse of BST was discussed by the Farm Animal Welfare Council and their recommendations for the production of guidelines in conjunction with the Royal College of Veterinary Surgeons should be accepted. They, however, did not cover two points. The first is, what happens if the farmer increases the recommended dose rate? The long term toxicology studies suggest that very little would happen to the cow. Such an action would be very expensive and soon discontinued. The second point is what happens if by accident the cow receives a massive dose of BST? Even when cows received their lifetime (4 lactations) dose of BST in a two week period, they remained healthy and subsequent slaughter and examination showed no abnormalities. (Vicini et al., 1988).

It has been suggested that since high producing cows are predisposed to production diseases (Erb, 1987 would disagree) which are increased by concentrates, any additional factor such as BST, which is associated with more concentrates (see Peel et al., 1987 for response to BST from grass alone) and higher production would inevitably cause a higher incidence of disease. The evidence does not truly support this hypothesis. Should we perhaps on these arguments stop breeding for higher yields through the selection of improved genetic stock? We should also remember that historically levels of mastitis have fallen as yield increased. This no doubt was linked to improved management techniques, something we can still look forward to. In addition it has been suggested that if the incidence of mastitis was increased, and that mastitis causes pain through pressure on inflamed tissues, BST would result in greater pain. This is another hypothesis difficult to sustain in the light of Professor Burvenich's work which suggests that BST may well help in the recovery from mastitis.

Bauman's comments made in (1987) are appropriate to quote. He said that "stressed animals would expend greater than normal energy (heat) for maintenance, produce less milk and have a lower productive efficiency. None of these stress related effects have been observed in cows treated with somatotropin."

Welfare Benefits?

We may conclude from the discussion so far that the proper use of BST does not pose a threat to animal welfare. BST undoubtedly represents a significant innovation in dairy husbandry. By taking a positive approach we may identify new opportunities for improving both productivity and animal welfare. Some possibilities to consider are:

1. Genetic selection of superior cows has inevitably led to selection for higher peak yields and consequently ever greater metabolic demands on cows during early lactation. An alternative approach would be to achieve equally high full lactation yields, but without adding to peak yields, by "turning up" production in later lactation with BST. This has been acknowledged in recent studies on the hormonal basis and effects of genetic selection for milk yield. Bonczek et. al., 1988 stated "Given the current controversy surrounding the use of BST, it is tempting to suggest that what might be achieved almost instantly by injecting BST could be achieved naturally by selection over a period of years. Further, the gradual changes brought about by selection would be much less disruptive to the market than would the more rapid change produced by injection of somatotropin. However, our results and those of Kazmer et. al., (1986) suggest that the increase in BST concentration achieved by selection does not persist totally throughout lactation. Hence while selection has been shown to increase plasma BST concentrations in early lactation, maintenance of high concentrations of BST throughout lactation may require BST injection to supplement selection."

2. The above approach would permit some redirection of genetic selection pressure towards health and/or conformation characteristics.

3. The possibility of prolonging useful lactations reduces the economic arguments for a 365 day calving interval. Current pressures on cows to return to breeding condition and conceive within 3 months of calving or be culled for infertility could be relaxed.

4. The potential of BST for restoring milk yield in cows following acute coliform mastitis, thereby avoiding culling for low production, is worthy of further exploration.

REFERENCES

Aguilar, A.A., Jordan, D.C., Olson, J.D., Bailey, C. & Hartnell, G. (1988). A short term study evaluating the galactopoietic effects of the administrtion of Sometribove (recombinant methionyl bovine somatotropin) in high producing dairy cows milked three times per day. Journal of Dairy Science, $\underline{71}$ (Supp) 208 (Abstract P269)

Allen, W.M. (1988) Personal Communication

Asimov, G.J. & Krouze, N.K. (1937). The lactogenic preparations from the anterior pituitary and the increase of milk yield of cows. Journal of Dairy Science, 20 289 - 306.

Bauman, D.E., Eppard, P.J. & McCutcheon (1984). Effects of exogenous somatotropin in lactating dairy cows. In New trends in animal nutrition and physiology. Monsanto International Symposium. Louvain-la-Neuve, Belgium.

Bauman, D.E. (1987). Bovine somatotropin. The Cornell experience. In National Invitational Workshop on Bovine Somatotropin. St. Louis Missouri Sept. 21-23

Bauman, D.E., Hard, D.L., Crooker, B.A., Erb, H.N. & Sandles, L.D. (1988). Lactational performance of dairy cows treated with a prolonged-release formulation of methionyl bovine somatotropin (sometribove). Journal of Dairy Science, $\underline{71}$ (supp 1) 205 (Abstract P260).

Bonczek, R.R., Young, C.W., Wheaton, J.E. & Miller, K.P. (1988). Response of somatotropin, insulin, prolaction and thyroxine to selection for milk yield in Holsteins. Journal of Dairy Science, $\underline{71}$ 2470-2479.

Booth, J.M. (1988). Progress in controlling mastitis in England and Wales. Veterinary Record, $\underline{122}$, 299-302

Bourchier, C.P., Garnsworth, P.C., Hutchinson, J.M. & Benson, T.A. (1987). The relationship between milk yield, body condition and reproductive performance in high yielding dairy cows. Animal Production $\underline{41}$ 460 (Abstract).

Burton, J.H., McBride, B.W., Bateman, K., MacLeod, G.K. Eggert, R.G. (1987). Recombinant bovine somatotropin : effects on production and reproduction in lactating cows. Journal of Dairy Science, $\underline{70}$ (Supp 1) 175 (Abstract P214).

Burvenich, C. (1988) Personal Communication

Chalupa, W., Baird, L., Soderholm, C., Palimquist, D.L. Hemken, R., Otterby, D., Annexstad, R., Vecchiarelli, B., Harman, R., Sinha, A., Linn, J., Hansen, W., Ehle, F., Schneider, P. & Eggert, R. (1987). Response of dairy cows to somatotropin. Journal of Dairy Science, $\underline{70}$ (Supp 1) 176 (Abstract P216)

Chalupa, W., Hutchens, A., Swager, D., Lehenbauer, T., Vecchiarelli, B., Shaver, R. & Robb, E. (1988). Response of cows in a commercial dairy to somatotropin. Journal of Dairy Science, $\underline{71}$ (Supp 1) 210, (Abstract P274).

Cole, W.J., Eppard, P.J., Lanza, G.M., Hintz, R.L., Madsen, K.S., Franson, S.E., White, T.C., Ribelin, W.E., Hammond, B.G., Bussen, S.C., Leak, R.K. & Metzger, L.E. (1988). Response of lactation dairy cows to multiple injections of sometribove, USAN (recombinant methionyl bovine somatotropin) in a prolonged release system. Part II. Health and reproduction. Journal of Dairy Science, $\underline{71}$, (Supp 1) 184 (Abstract P196).

Dhiman, T.R., Kleinmans, J., Radloff, H.D., Tessmann, N.J. and Satter, L.D. (1988). Effect of recombinant bovine somatotropin on feed intake, dry matter digestibility and blood constituants in lactating dairy cows. Journal of Dairy Science, 71 (Supp 1), 121 (Abstract P8).

Ducker, M.J., Haggett, R.A., Fisher, W.J., Morant, S.V. & Bloomfield, G.A. (1985). Nutrition and reproductive performance of dairy cattle. Animal Production 41 1-12.

Elvinger, F., Head, H.H., Wilcox, C.J. & Natzke, R.P. (1987). Effects of administration of bovine somatotropin on lactation milk yield and composition. Journal of Dairy Science, 70 (Supp) 121, (Abstract P52).

Eppard, P.J., Bauman, D.E., Curtis, C.R., Erb, H.N., Lanza, G.M. & DeGeeter (1987). Effect of 188 day treatment with somatotropin on health and reproductive performance of lactating dairy cows. Journal of Dairy Science 70 582-591.

Erb, H.N. (1987). Interrelationship among production and clinical disease in dairy cattle : A review Canadian Veterinary Journal, 28 326-329.

Esslemont, R.J., Baile, J.H. & Cooper, M.J. (1985) Infertility management in dairy cattle. Publishers Collins London.

Furniss, S.J., Stroud, A.J., Brown, A.C.G. & Smith, C. (1988). Milk production, feed intakes and weight change of autumn calving, flat rate fed dairy cows given two-weekly injections of recombinantly derived bovine somatotropin (BST). Animal Production 46 483 (Abstract).

Gluckman, P.D., Breier, B.H. & Davis, S.R. (1987). Physiology of the somatototropic axis with particular reference to the ruminant. Journal of Dairy Science, 70 442-466.

Hard, D.L., Cole, W.J., Franson, S.E., Samuels, W.A., Bauman, D.E., Erb, H.N., Huber, J.T. & Lamb, R.C. (1988). Effect of long term sometribove treatment in a prolonged release system on milk yield, animal health and reproductive performance pooled across four sites. Journal of Dairy Science, 71 (Supp 1) 210 (Abstract P273).

Hemken, R.W. Hamon, R.J. Silvia, W.J., Heersche, G. & Eggart, R.G. (1988). Response of lactating dairy cows to a second year of recombinant bovine somatotropin (BST) when fed two energy concentrations. Journal of Dairy Science, 71 (Supp 1) 122 (Abstract P9).

Huber, J.T., Willman, S., Marcus, K., Theurer, C.B., Hard, D.L. & Kung, L. (1988). Effect of sometribove, USAN (recombinant methionyl bovine somaototropin) injected in lactating cows at 14-d intervals on milk yield, milk composition and health. Journal of Dairy Science, 71 (Supp 1) 207 (Abstract P264).

Hutchison, C.F., Tomlinson, J.E. & McGer, W.H. (1986). The effects of exogenous recombinant or pituitary extracted bovine growth hormone on the performance of dairy cows. Journal of Dairy Science, 69 (Supp 1) 152 (Astract P185)

Huth, F.W., & Schutzbar, W.V. (1987). Einflus der Milchleistung auf die Zwischenkalbezeit. Deutsche Schwarzbunte 11, 12-14.

Kazmer, G.W., Barnes, M.A., Akers, R.M. & Pearson, R.E. (1986). Efect of genetic selection for milk yield and increased milking frequency on plasma growth hormone and prolactin concentration in Holstein cows. Journal of Animal Science, 63 1220-1227.

Kronfeld, D.S., (1987). Health risks in dairy cows given biosynthetic somatotropin. Nutrition Institute Proceedings, National Feed Ingredients Association, West Des Moines, Iowa

Lamb, R.C., Anderson, M.J., Henderson, S.L., Call, J.W., Callan, R.J., Hard, D.L. & Kung, L. (1988). Production response of Holstein cows to sometribove, USAN (recombinant methionyl bovine somatotropin) in a prolonged release system for one lactation. Journal of Dairy Science, 71, (Supp 1) 208, (Abstract P267).

Lanza, G.M., White, T.C., Dyer, S.E., Hudson, S., Franson, S.E., Hintz, R.L., Duque, J.A., Bussen, S.C., Leak, R.K. & Metzger, L.E. (1988). Response of lactating dairy cows to intramuscular or subcutaneous injections of sometribove, USAN (recombinant methionyl bovine somatotropin) in a prolonged release system. Part II changes in circulating analyses. Journal of Dairy Science, 71 (Supp) 195 (Abstract P228).

McGuffey, R.K., Green, H.B. & Ferguson, T.H. (1987). Lactation performance of dairy cows receiving recombinant bovine somatotropin by daily injection or in a sustained release vehicle. Journal of Dairy Science, 70 (Supp) 176 (Abstract P218).

McGuffey, R.K., Green, H.B. & Basson, R.P. (1988). Protein nutrition of the somatotropin treated cow in early lactation. Journal of Dairy Science, 71 (Supp 1) 120 (Abstract P5).

Mollett, T.A., DeGeeter, M.J. Belyea, R.L., Youngquist, R.A. & Lanza, G.M. (1986). Biosynthetic or pituitary extracted bovine growth induced galactopoiesis in dairy cows. Journal of Dairy Science, 69 (Supp 1) 118 (Abstract P83).

Monsanto, plc (1988) Personal Communication

Munneke, R.L., Sommerfeldt, J.L. & Ludens, E.A. (1988). Lactational response of dairy cows to recombinant bovine somatotropin. Journal of Dairy Science 71 (Supp 1) 206 (Abstract P263).

Palmquist, D.L. (1988). Response of high producing cows given daily injections of recombinant bovine somatotropin from D30-296 of lactation. Journal of Dairy Science, 71 (Supp) 206 (Abstract P261).

Peel, C.J. (1988). Bovine somatotropin (BST) - A review of efficacy and mechanism of action in dairy cows. European Association of Veterinary Pharmacology and Toxicology. Budapest September 1st 1988.

Peel, C.J., Sandles, L.D, Quelch, K.T. & Herington, A.C. (1985). The effects of long term administration of bovine growth hormone on the lactational performance of identical twin dairy cows. Animal Production 41 135-142.

Peel, C.J., Eppard, P.J. & Hard, D.L. (1988). Evaluation of sometribove (methyional bovine somatotropin) in toxicology and clinical trials in Europe and the United States. Presented at International Symposium on Biotechnology in Growth Regulation. 18-20 Sept. Cambridge U.K.

Peel, C.J., & Bauman, D.E. (1988). Somatotropins and lactation. Journal of Dairy Science, 70 474-486.

Pell, A.N. Tsang, D.S. Huyler, M.T., Howlett, B.A., Kunkel, J. & Samuels, W.A. (1988). Response of Jersey cows to treatment with sometribove, USAN (recombinant methionyl bovine somatotropin) in a prolonged release system. Journal of Dairy Science, 71 (Supp 1) 206 (Abstract P262).

Phipps, R.H., Weller, R.F., Austin, A.R., Craven, N. & Peel, C.J. (1988). A preliminary report on a prolonged release formulation of bovine somatotropin with particular reference to animal health. Veterinary Record 512-513.

Phipps, R.H. (1988). The use of prolonged release somatotropin in milk production. International Dairy Federation Bulletin No. 228.

Phipps, R.H. (1988). Unpublished data.

Phipps, R.H., & Weller, R.F. (1988). Efficacy data for British Friesians treated for two consecutive lactations with a prolonged release formulation of bovine somatotropin. (To be presented at XV World Buiatrics Conference, Palma, 11-14 Oct. 1988).

Rowe-Bechtel, C.L. Muller, L.D., Deaver, D.R. & Griel, L.C. (1988). Administration of recombinant bovine somatotropin (rbst) to lactating dairy cows beginning at 35 and 70 days postpartum. I Production response. Journal of Dairy Science, 71 (Supp 1) 166, (Abstract P142).

Rijpkema, Y.S. Reeuwyk, L. van., Peel, C.J. & Mol, E.P. (1987). Response of dairy cows to long-term treatment with somatotropin in a prolonged release formulation. 38th Annual Meeting of the European Association for Animal Production.

Soderholm, C.G., Otterby, D.E., Linn, J.G., Ehle, F.R., Wheaton, J.E., Hansen, W.P. & Annexstad, R.J. (1988). Effects of recombinant bovine somatotropin on milk production, body composition and physiological parameters. Journal of Dairy Science, 71 355-365.

Soderholm, C.G., Otterby, D.E., Linn, J.G., Wheaton, J.E., Hansen, W.P. & Annexstad, R.J. (1986). Effects of different doses of recombinant bovine somatotropin on circulating metabolites, hormones and physiological parameters in lactating cows. Journal of Dairy Science, 69 (Supp 1) 152 (Abstract P183).

Tessman, N.J. Kleinmans, J., Dhiman, T.R. Radloff, H.D. & Satter, L.D. (1988). Effect of dietary forage: grain ratio on response of lactating dairy cows to recombinant bovine somatotropin. Journal of Dairy Science, 71 (Supp 1) 121, (Abstract P7).

Thomas, C., Johnsson, I.D., Fisher, W.J., Bloomfield, G.A., and Morant, S.V. (1987). Effect of recombinant bovine somatotropin in milk production reproduction and health of dairy cows. Animal Production 44 460 (Abstract)

Vicini, J.L., De Leon, J.M. Cole, W.J., Eppard, P.J., Lanza, G.M., Hudson, S., and Miller, M.A. (1988). Journal Dairy Science 71 (Supp. 1), 168 (Abstract P147).

Webster, A.J.F. (1985). Farm Animal Welfare. The needs of animals and the wishes of society. British Society of Animal Production : Winter Meeting Paper No. 39.

Whitaker, D.A., Smith, E.J., Kelly, J.M. & Hodgson-Jones, L.S. (1988). Health welfare and fertility implications of the use of bovine somatotropin in dairy cattle. Veterinary Record 503-505.

Wollny, C. (1988) Personal Communication.

INFLUENCES OF SOMATOTROPIN ON EVALUATION OF GENETIC MERIT FOR MILK PRODUCTION

H. O. Gravert
Federal Dairy Research Center,
P. O. Box 60 69, D - 2300 Kiel 14

ABSTRACT

BST might affect the evaluation of genetic merit and the selection response if there is an interaction between genotype and BST-response, if production figures are manipulated and if wrong animals are selected. From the BST-experiments there is no indication for a genotype-response interaction, neither from breed comparisons nor from twin experiments or the repeatability of BST-response in consecutive lactations. The cow evaluation might be biased by administration of BST, however, this can be corrected for when the BST administration is recorded by the milk recording service. BST will affect the bull evaluation only slightly if BST is administered randomly. The genetic progress in milk yield will be nearly unchanged because BST-application will not change the ranking of bulls and cows dramatically.

INTRODUCTION

In connection with the BST-discussion farmers and breeders are particularly interested in the breeding aspects of BST-administration. Their opinion ranges from mild scepticism to absolute reluctance. Mostly the discussion centers on three questions:

(1) Is there a relationship between genetic merit and response on BST-treatment?
(2) How would BST-treatment affect
    - cow indexes,
    - bull indexes, and
    - genetic progress?
(3) What would be the consequences for the breeding policy and the breeder organizations?

1. COW-BST-INTERACTION

Coming to the first question about genetic merit and BST-response one might speculate about the role of somatotropin within high yielding cows. It is well known that the level of BST in blood is related to the milk yield, as shown by Hart et al. (1978, 1979, 1980), Bines et al. (1982) or Kazmer et al. (1986). Therefore one could expect that the BST response would be different in low and high yielding cows and also between breeds with low and high milk potential. Apparently this is not the case (table 1). Though there have been very few experiments comparing different breeds and with small numbers like that of Oldenbroek et al. (1987) we can not detect any larger response in breeds with lower milk yields. In high yielding Holsteins the superiority of BST-treated cows has been around 4 - 5 kg FCM or 17 - 19 %, in Jersey there was about the same absolute increase which corresponds to a higher relative increase (23 - 33 %), and in low yielding Fleckvieh- and Normande-cows the superiority was about 2 - 3 kg and 15 - 22 % resp. Therefore in total we get the impression that the BST-response is approximately proportional to the breed average.

Within breeds the picture is also not very clear (table 2). Peel et al. (1988) did not observe any relationship between base milk yield and lactation response to BST for individual cows in four European BST-trials with 102 treated cows. The numbers in other experiments were smaller, however, they agree that there seemed to be no or a slightly negative relationship, i. e. low yielding cows showed a higher response to BST-treatment than cows with a higher base yield, though the relationship was rather weak and not significant. In our own data on 64 cows the correlation was $r = -0.16$.

Another indication for individual reaction on BST-treatment is the repeatability of the response from one year to another. Unfortunately very little has been published on this question. From our own data we found a repeatability of $r = 0.20$ for 20 cows with two lactations, however, the repeatability in the control group was even higher ($r = 0.38$), indicating that cows have similar lactation curves in consecutive years. BST-treatment shifted the curve upwards without any special cow-treatment-interaction.

Another approach to answer the first question is to look at identical twins. We included 6 pairs of monozygous twins into our BST-trial, one cow of each pair was treated. If the BST-response would be different between genotypes then we would have to expect different responses between MZ pairs. Due to the high daily variation in milk yield, even during the pre-treatment period, we were not able to distinguish different reactions.

If there is no repeatability for BST-response and no difference between MZ-pairs then it is unlikely that there exists a relationship between genetic merit and response on BST administration. As an example McDaniel and Hayes (1988) treated 204 cows with four levels of BST and regressed the milk yields on PDs of their sires. The F-value for interaction of PD milk and dose effect was 1.1. All evidence suggested high yielding cows and daughters of bulls with high PDs will respond to BST equally as well as cows of lower merit. In our own data the correlation between BST-response and breeding value of the sire was $r = 0.11$ which confirms the general statement that there is no relationship between genetic merit and reaction on BST-treatment.

Therefore we come to the conclusion that BST would be a management tool independent of genetic progress. It would increase the level of production like better feeding and management. However, will BST increase the variability of milk yields? Yes, if we look at the day-to-day variation of an individual cow because there is a cyclic pattern of the milk yield between the days of injection. This means that the releasing system must be improved. The answer is No, if we look at the standard deviation between cows for the average milk yield within an injection-cycle of two weeks (table 4). In our trial the standard deviations of the average daily milk yield per cow during the two weeks before first BST injection were

6,5 and 7,5 kg for the control and treatment group resp. During BST administration the standard deviations were similar and decreased slightly with the progress of lactation.

2. BST AND COW EVALUATION

The answer to the second question: how would BST-treatment affect cow indexes, will depend on the model of the cow index and on the strategy of BST administration. In most countries the cow index is the basic information to select potential bull dams and it includes several traits like milk yield, fat and protein content, body and udder conformation, milkability etc. In addition to the cow's own performance the genetic merit of relatives, mostly from her sire, is included into the cow index. Therefore improvements in one trait like the cow's own milk yield will change her index to a minor extent. The relative weights given to all information available to calculate the cow index rely on heritabilities, correlations and economic terms. With random use of BST we would expect a reduction of the heritability for milk yield, say from about $h^2 = 0.25$ to 0.16 at maximum. Therefore less weight would be put on the cow's own performance and more on the breeding value of her sire.

The most uncertain issue will be the strategy of BST application. In my opinion most farmers would use it only temporarily
- to produce more milk when the milk prices are high
- to adjust their milk production to a given quota and
- to avoid fat cows in the second half of lactation.

Nevertheless let us assume that breeders will use BST to achieve higher production figures for potential bull dams. A simple example might clarify the situation: if the minimum requirement for a cow index of a bull dam would correspond to a production level of 8000 kg milk, without BST 8,4 % of the German Friesian cows would be eligible (table 5). If all cows would be treated the ranking would be the same, the minimum requirement would be higher, say 9000 kg. If only high yielding cows would be treated which fulfill the minimum requirements already without BST, nothing would be changed. If on the other hand only cows with less than 7000 kg would be treated they would even with BST not become eligible as bull dams. So there is only a certain group of cows (16,2 %) where BST treatment would make the cows eligible. However, if then 24,6 % of all cows fullfil the minimum requirement of 8000 kg milk and just 8,4 % are needed as potential bull dams the minimum requirement might be raised and the truncation point will be around 8500 kg. Then about one half of the potential bull dams are classified incorrectly and overestimated. In combination with the sires breeding value and other traits the proportion of incorrect ranking will be less.

3. BST AND BULL EVALUATION

The effect of BST-treatment on the accuracy of estimating genetic merit of bulls has been simulated in computer models (Colleau, 1988, Burnside and Meyer, 1987, Leitch, 1986, Everett, 1987, Frangione and Cady, 1988, Simianer and Wollny, 1988). The results agree that there will be a slight reduction in the accuracy of bull evaluation (table 6). The reduction will reach its maximum, if 50 % of the cows are treated, if the BST effect is large and if there is a cow x BST-interaction. But even in this case it will be less than 10 % and it could be counterbalanced by a larger number of daughters per sire. The ranking of bulls will change very little and without any practical significance.

The situation may be different if the administration of BST is not randomly distributed. We might think of "syndicate-bulls", owned by a group of breeders who administer BST to daughters of special bulls only. So the important question arises: can we detect records from BST-treated cows? From our own data with 60 cows Weber (1988) calculated a set of different discriminant functions. If we look for instance at the production figures during two weeks before and two weeks after first BST injection, the multiple correlation for the corresponding discriminant function (table 7) was $R = 0.86^{xxx}$. The discriminant power was mainly based on milk-yield and protein content. Based on this function 93 % of the untreated cows could be classified correctly but just 87 % of the treated cows. The other 17 % may be called "non-reactors" to BST since their production remained within the normal range. However, when we applied the same discriminant function to an independent sample in the second year, only 37 % of the treated cows were detected. For the practical use of such a function it means that it is not very powerful to detect BST administration. Nevertheless an unexpected increase in milk yield from one recording day to the next and a simultaneous decline in protein content will at least arouse the suspicion of a BST administration.

In my opinion each breeder should accept and sign an obligation to report each BST treatment to the official milk recorder at the test day. This report should include the date of injection and dose of BST. These figures will then allow generation of adjustment factors to correct for the BST effect before milk records are utilized for the estimation of genetic merit of bulls and cows. The phenotypic milk yields should not be corrected because these are real production figures. I do not see any reason to mistrust the breeders in general. Finally it would also be to their disadvantage if wrong animals are selected for breeding.

4. BST AND GENETIC PROGRESS

How would BST affect genetic progress? This question has two aspects: (1) what role would BST play within current breeding programs and (2) how far will BST shift the selection pressure to other traits than milk?

The current breeding programs are based on genetic parameters like heritability for milk yield of about $h^2 = 0.25$. Due to unknown or partly unknown use of BST the heritability will decrease. This can be compensated in progeny testing for bulls by more daughters per bull. It can not be compensated in the evaluation of bull dams. If we assume that the best 10 % of progeny tested bulls are used as sires of the next bull generation and that the repeatability of the test is 0.85, then the expected superiority of these sires will be + 1.62 genetic standard units (table 8). The breeding value of the dam is based on her own performance and on the breeding value of her sire and is about 0.40 under current conditions.

If the heritability of her own performance is reduced and the accuracy of the sire's breeding value is unchanged, then the repeatability of the cow's breeding value is about 0.35. With the same selection intensity of 3 % the genetic superiority of the bull dams will be reduced from 1.43 to 1.34 standard units. Therefore average young bulls will have 3 % lower breeding values. This can be compensated for by a few more young test bulls.

In modern breeding schemes with use of embryo transfer, e. g. MOET, unknown BST administration might be more serious if donor cows are manipulated. In this case the supervision of donor cows should be intensified, e. g. by weekly milk recording.

For the second part of the question we can assume that any new development of production enhancers (like BST) will change the emphasis put on different traits. We know for instance that selection for higher milk yield has mainly improved the initial yield during the first part of lactation because its genetic variance is larger than for the daily milk yield at later stages of lactation. High yielding cows suffer during the post-partum energy deficit with well-known effects on fertility, occurrence of ketosis and low protein content in milk. Milk yield and persistency are negatively correlated. BST will allow to put less weight on initial yield and to improve the persistency.

If less selection pressure must be put on milk yield more weight can be given to resistance against mastitis, to better temperament, to better legs and hooves etc. This will enhance the genetic progress in these secondary but important traits.

5. BST AND BREEDER ORGANIZATIONS

Breeder organizations feel that BST might become an indirect threat to their activities if progress can be achieved by synthetic means also. On the other hand they do not want to stay behind any modern development like some of them did when artificial insemination was introduced. Sales of semen and breeding animals has become the back-bone of to-days breeder organizations and competition on the international market is strong. Therefore the attitude towards BST administration has already been discussed at the World Holstein Friesian Conference this year at Nairobi. It was agreed that the breeder organizations will not promote BST but its administration will not be prohibited for herdbook members. If we look at BST as a new management tool it will not change the need for better livestock nor will it affect the breeder organizations as such. Whether the breeder organizations feel it nessessary to install test stations for bull dams, will depend more on the general structure of the organization than on the possible administration of BST. It is agreed, however, that BST will open new opportunities to manipulate phenotypic production records and that ways have to be found to avoid any negative effects on the efficiency of modern breeding schemes.

TABLE 1   BST-response in different breeds

author	breed	response kg milk	%
Oldenbroek et al., 1987	Friesians	4,2	17
	Red and White	3,0	14
	Jersey	3,7	23
Chalupa et al., 1987	Holstein	4,9	18
	Jersey Brown Swiss Ayrshire	5,8	24
Peel et al., 1988	Friesians	4,1	19
Kräußlich, 1988	Fleckvieh	2,3	15
Lossouarn, 1988	Normands	3,0	22
Pell et al., 1988	Jersey	5,6	33

TABLE 2 Relationship between base milk yield and BST-response

author	n	relationship
Peel et al., 1988	102	none
Leitch et al., 1987	34	negative
Oldenbroek et al., 1987	35	negative
Rohr, 1987	?	negative
Huber et al., 1988	80	none
Eppard et al., 1987	30	none

TABLE 3 Relationship between genetic merit and BST-response

author	n	relationship
Nytes, et al., 1988	39	none
McDaniel et al., 1988	204	none
Leitch et al., 1987	34	none
Weber, 1988	30	r = 0.11

TABLE 4 Standarddeviations of average milk yield per cycle (two weeks) between cows

| group | n | Pre-treatment | cycle | | | | | | | | 1-8 |
			1	2	3	4	5	6	7	8	
control	30	6,5	6,6	6,3	6,0	5,8	5,8	5,7	5,7	5,6	5,9
BST	30	7,5	7,1	7,3	7,0	7,0	7,3	7,2	6,8	6,7	7,1

TABLE 5 Distribution of milk yield without and with BST (BST effect = + 1000 kg)

BST application	-6000	6000-7000	7000-8000	8000-9000	9000-10000	10 000-
none	47,5	27,9	16,2	6,2	1,7	0,5
all cows		47,5	27,9	16,2	6,2	2,2
only above 8000 kg	47,5	27,9	16,2	-	6,2	2,2
only below 7000 kg		47,5	27,9	6,2	1,7	0,5
7000 - 8000	47,5	27,9	-	22,4	1,7	0,5

TABLE 6 The accuracy of sire evaluation (r_{AI}) for different effects of BST on daily milk yield and percentage of cows treated (from Simianer and Wollny, 1988)

Effect of BST (kg)	BST x cow interaction variance (kg^2)	% of cows treated		
		0	20	40
2	0.25	.74	.74	.74
4	1.00		.74	.74
6	2.25		.73	.73
8	4.00		.73	.71

TABLE 7 Discriminant function with five traits before and after BST injection (Weber, 1988)

	milk-kg	fat-%	protein-%	lactose-%	SCC
before BST	− 4.10	.11	1.91	− .63	− .21
after BST	+ 4.08	.36	− 2.13	.34	.49

	correct classified (%)	
	without BST	with BST
in same sample	93	87
in another sample	90	37

TABLE 8 Effect of BST on breeding values of young bulls (in σ_A)

	without BST		with BST	
	repeatability	breed.v.	repeatability	breed.v.
bull sire	0.85	1.62	0.85	1.62
bull dam	0.40	1.43	0.35	1.34
young bulls	0.22	1.53	0.21	1.48

REFERENCES

Bines, J.A. and Hart, I.C. 1982: Metabolic limits to milk production, especially roles of growth hormone and insulin. J. Dairy Sci., 65, 1375-1389.

Burnside, E.B. and Meyer, K. 1987: Effect of different strategies of administration of bovine somatotropin on within herd variance and accuracy of sire rankings for milk production. J. Dairy Sci., 70, Suppl. 1, 127.

Chalupa, W., Baird, L., Soderholm, C., Palmquist, D.L., Hermken, R., Otterby, D., Annexstad, R., Vecciarelli, B., Harmon, R., Sinha, A., Linn, J., Hansen, W., Ehle, F., Schneider, P. and Eggert, R. 1987: Responses of dairy cows to somatotropin. J. Dairy Sci., 70, Suppl. 1, 176.

Colleau, J.J. 1988: pers. comm.

Eppard, P.J., Bauman, D.E., Curtis, C.R., Erb., H.N., Lanza, G.M. and Degeeter, M.J. 1987: Effect of 188-day treatment with somatotropin on health and reproductive performance of lactating dairy cows. J. Dairy Sci., 70, 582591.

Everett, R.W. 1987: How will BST affect dairy genetics in the 1990's? Hoards Dairyman, April 10.

Frangione, T. and Cady, R.A. 1988: Effects of bovine somatotropin on sire summaries for milk production and milk yield heritabilities. J. Dairy Sci., 71, Suppl. 1, 239.

Hart, I.C., Bindes, J.A., Morant, S.V. and Ridley, J.L. 1978: Endocrine control of the levels of hormones (prolactin, growth hormone, thyroxine and insulin) and metabolites in the plasma of high and low yielding cattle at various stages of lactation. J. Endocrinol., 77, 333-345.

Hart, I.C., Bines, J.A. and Morant, S.V. 1979: Endocrine control of energy metabolism in the cow: correlations of hormones and metabolites in high and low yielding cows for stages of lactation. J. Dairy Sci., 62, 270-277.

Hart, I.C., Bines, J.A. and Morant, S.V. 1980: The secretion and metabolic clearance rates of growth hormone, insulin and prolactin in high and low yielding cattle at four stages of lactation. Life Sci., 27, 1839.

Huber, J.T., Franson, S.E., Hoffman, R.G. and Hard, D.L. 1988: Relationship of production level and days postpartum to response of cows to sometribove. J. Dairy Sci., 71, Suppl. 1, 207.

Kazmer, G.W., Barnes, M.A., Akers, R.M. and Pearson, R.E. 1986: Effect of genetic selelection for milk yield and increased milking frequency on plasma growth hormone and prolactin concentration in Holstein cows. J. Anim Sci., 63, 1220-1227

Kräußlich, H., 1988: pers. comm.

Leitch. H.W., 1986: Genetic and phenotypic aspects of administration of recombinant bovine somatotropin to Holstein cows. M. S. Thesis, Guelph, Canada.

Leitch, H.W., Burnside, E.B., MacLeod, G.K., McBride, B.W., kennedy, B.W., Wilton, J.W. and Burton, J.H. 1987: Genetic and phenotypic affects of administration of recombinant bovine somatotropin to Holstein cows. J. Dairy Sci., 70, Suppl. 1, 128.

Lossouarn, C., 1988: pers. comm.

McDaniel, B.T. and Hayes, P.W. 1988: Absence of interaction of merit for milk with recombinant bovine somatotropin. J. Dairy Sci., 71, Suppl. 1, 240.

Nytes, A.J., Combs, D.K. and Shook, G.E. 1988: Efficacy of recombinant bovine somatotropin injected at three dosage levels in lactating dairy cows of different genetic potentials. J. Dairy Sci., 71, Suppl. 1, 123.

Oldenbroek, J.K., Garssen, G.J., Forbes, A.B. and Jonker, L.J. 1987: The effect of treatment of dairy cows of different breeds with recombinantly derived bovine somatotropin in a sustained delivery vehicle. 38th Annual Meet. EAAP, Lisbon, N 5.

Peel, C.J., de Kerchove, G., Schockmel, L.R. and Craven, N. 1988: Recent developments in use of somatotropins in lactating dairy cows. Proc. VI World Conf. Animal Prod., Helsinki, 391.

Pell, A.N., Tsang, D.S., Huyler, M.T., Howlett, B.A. and Kunkel, J. 1988: Responses of Jersey cows to treatment with sometribove in a prolonged release system. J. Dairy Sci., 71, Suppl. 1, 206.

Rohr, K., 1987: Milchleistung und Milchzusammensetzung. BST-Symposium Völkenrode 3.-4.11.1987, 104-115.

Simianer, H. and Wollny, C., 1988: Impact of the potential use of bovine somatotropin on the accuracy of sire selection. Polycopy, 21 p.

Weber, T., 1988: Wirkung von rekombinantem bovinen Somatotropin (BST) bei Milchkühen in zwei aufeinanderfolgenden Laktationen. Diss. Kiel, 152 p.

EFFECTS OF ADMINISTRATION OF SOMATOTROPIN ON
GROWTH, FEED EFFICIENCY AND CARCASS COMPOSITION
OF RUMINANTS: A REVIEW.

W. J. Enright*

Teagasc, The Agriculture and Food
Development Authority.
* Grange Research Centre, Dunsany, Co. Meath,
Ireland.

ABSTRACT

Somatotropin (growth hormone; GH) of pituitary origin is necessary for normal growth of farm animal species. Growth hormone is a protein which can be produced in large quantities by recombinant DNA technology. For these reasons, exogenous GH has potential to be an effective commercial growth or carcass enhancer. This paper summarizes the available evidence on the effects of exogenous GH and GH-releasing factor on growth, feed efficiency and carcass composition in cattle and sheep. An insufficient number of long term studies have been performed and published to date. Many important basic and applied experiments still need to be performed. For example, the effects of dose, duration, frequency of administration and dietary regimes on response to GH need to be more fully addressed. From the available evidence, administration of GH to cattle and sheep increases growth, feed efficiency and carcass lean, while decreasing carcass fat. Thus, GH appears to satisfy the commercial requirements of an effective growth/carcass enhancer for ruminants.

INTRODUCTION

Net income to farmers from livestock production is directly influenced by body weight gain, feed efficiency and carcass composition. In addition, the demand for meat, especially beef and sheep meat, by consumers is increasingly influenced by the fat content of meat (Kempster, 1988). It is publicly recognised that in humans there is a causal link between dietary fat intake and development of cardiovascular and other diseases (Anonymous, 1984). Furthermore, the production of trimmable fat in the meat industry is excessive

(Kempster et al., 1986) and unwanted, and is energetically very inefficient relative to muscle (Van Es, 1977). To benefit farmers, the meat industry and consumers, research efforts should be directed towards the production of leaner carcasses. A secondary goal of research should be to produce these leaner carcasses more efficiently.

Leaner carcasses can be achieved in several ways including the use of "lean" breeds, within - breed selection, using males versus females, using intact versus castrate animals, slaughtering at a lighter weight, using low energy-high protein dietary regimes, long-day versus short-day photoperiod and various physiological manipulations. For a particular producer many of the above may not be attractive for economic, management or other reasons.

Within the sphere of physiological manipulations, the use of gonadal steroid implants, primarily in cattle, are particularly attractive (Roche and Quirke, 1986) and have been and are used worldwide. However, approximately two years ago the use of steroids was prohibited in European Community countries (Directive 85/649 EEC) in spite of (1) the large amount of scientific data demonstrating the efficacy and safety of the natural steroids (Lamming et al., 1987), (2) the fact that steroid implants are still being used by many of the EC trading partners and (3) approval of their use by the Food and Agriculture Organisation/World Health Organisation Expert Committee. Other physiological manipulations include the use of pituitary hormones eg. growth hormone (GH); hypothalamic hormones eg. growth hormone-releasing factor (GRF); other hormones/factors/ modified amino acids eg. somatomedins and beta-agonists; immunization eg. against somatostatin (SRIF) and fat cell membranes; and transgenic animals with extra copies of the gene for GH. All of these methods are presently being researched. Several methods (i.e., GH, GRF, transgenic animals) are based on the hypothesis that elevated concentrations of GH in the blood will be translated into an improvement in growth, efficiency and (or) carcass leanness. The use of transgenic animals is still a long term prospect in animal production while the use of exogenous GH or GRF to commercially enhance growth, efficiency and carcass leanness has considerable potential. With the advent of recombinant DNA technology, GH and GRF can be economically produced in large quantities for commercial use. Because of their galactopoietic properties there has been considerable research activity recently with these compounds in lactating cows and adequate delivery systems are close to completion.

This paper will review all the available data, published and unpublished, on the effects of administration of GH and GRF on growth, feed efficiency and carcass composition in cattle and sheep. No such information has been found for the goat.

Tables 1 and 2 provide a summary of materials, methods and results for the studies in cattle and sheep, respectively, which have examined the effects of exogenous GH on performance and carcass characteristics. Note that studies vary in many respects such as source of GH (pituitary or recombinant; ovine or bovine); dose and biological potency of GH preparations; route and frequency of administration; number, sex, type and age of animals; duration of treatment; level and type of nutrition; and the specific methodology used to determine the characteristics measured. With this in mind, one must take caution when comparing studies and summarizing overall effects. Nonetheless, an attempt will be made to draw some conclusions from the available data.

CATTLE
Growth rate

Brumby (1959) was the first to examine the effect of GH on growth. Six young Jersey heifers received daily subcutaneous injections of pituitary GH [110 ug/kg body weight (BW)] for 84 days which significantly improved average daily gain (ADG) approximately 10% above controls. However, the GH preparation used contained substantial quantities of thyrotrophin - stimulating hormone which may have influenced (probably adversely) the growth response to GH per se. In addition to ADG, GH treatment significantly increased wither height approximately 16% during the treatment period. Two

TABLE 1   The effects of exogenous growth hormone on growth and carcass characteristics of cattle[a].

Reference	Treatment	Daily dose (ug/kg BW)	No. per trt.	Duration of trt.,d (period analysed)	Anim type, initial wt./age	Effect relative to controls (%)					Carcass			
						ADG	ADFI	FCE	KO%	Wt.	Lean	Fat	N ret.	
Brumby (1959)	10mg pit.GH (sc 1x/d)	110	6[b]	84 (1 to 84)	Heifers 89kg	+10* (.75)	N.D.	N.D.	N.D.	N.D.	N.D.	N.D.	N.D.	
Car et al. (1967)	140mg pit.GH (1m 1x/d)	376	4	25 (-8 to 36)	Steers 372kg 13mo	-17 (1.1)	None	-14[f]	N.D.	N.D.	N.D.	N.D.	+58* (31.9)	
zNidar (1976)	158mg pit.CH (1m 1x/d)	363	6	19 (-4 to 24)	Bulls 435kg 13.5mo	+6.5 (.61)	None	+8[c]	N.D.	N.D.	N.D.	N.D.	+81* (16.7)	
Moseley et al. (1982)	9.4mg iv cont. pit.GH/d iv 6x/d combined	48	8	10 (7 to 10)	Steers 192kg	N.D.	N.D.	N.D.	N.D.	N.D.	N.D.	N.D.	+21* +12* +15*	
Sejrsen et al. (1983, pers.comm.)	18mg pit.GH (inj. 1x/d)	100	9[b]	109 (1 to 98)	Heifers 179kg 8mo	+8.6* (.88)	None (Pair -fed)	+8.6*	0 (49)	+1.7 (134)	+2.2*	-12*	N.D.	
Wolfrom and Ivy (1985)	pit.GH 18mg (sc 1x/d) 36mg	50 100	9	8 (1 to 8)	Steers 360kg	+117 +151	N.D.	+46 +54	N.D.	N.D.	N.D.	N.D.	N.D.	
	pit.GH 6mg (sc 1x/d) 12mg 18mg	19 38 56	12		Steers 320kg	N.S. N.S. N.S.		N.S. N.S. N.S.						
Quirke et al. (1985, pers.comm.)	10.3mg pit.GH (sc 1x/d)	40	15	154 (1 to 154)	Steers 237kg 9mo	+6* (.95)	+3	+4*	+1[d] (54)	+3 (241)	N.D.	N.D.	N.D.	
Ivy et al. (1986)	pit.GH 6mg (sc 1x/d) 12mg 24mg	17 34 68	8	42 (1 to 42)	Steers 356kg	+* +* +*	N.D.	+* + +	N.D.	N.D.	N.D.	N.D.	N.D.	
Peters (1986)	20mg pit.GH (sc 1x/d)	74	6	29 (1 to 29)	Steers 270kg 10mo	+20 -6 (.26,2.3)	O(R) -14(AL)	+1 +1 (71,69 X 204,230)	0 -2	+10 +22*	-7 -32*	N.D.		
Eisemann et al. (1986)	22.5mg pit.GH (sc 1x/d)	61	6	14 (8 to 14)	Heifers 368kg 12mo	N.D.	+6 (N=8; d126 to 140)	+2 (N=6; d126 to 140)	0 (44) (N=4)	+9* (69) (N=4)	0 (N=4)	N.D.	+500* (2)	
Sandles and Peel (1987)	54mg pit.GH (sc 1x/d)	600	12[b]	147 (1 to 147)	Heifers 90kg 3.5mo	+8.6* (.49)					0 (N=4)	-17 (N=4)	+2.3 (20.6) (N=6;d126 to 140)	

TABLE 1 (continued).

Reference	Treatment												
Williams et al. (1987)	3.5mg GH (sc 1x/d)	70	5	100 (1 to 100)	Bulls .75mo	+3 (.98)	N.D.	N.D.	N.D.	N.D.	N.D.	N.D.	N.D.
Fabry et al. (1987)	22mg pit.GH (sc 1x/d)	50	10	126 (1 to 126)	Heifers 439kg 16mo	+23* (.93)	+1.5	+20*	-2 (58.9)	+3 (324)	+2	-3	N.D.
Kirchgessner et al. (1987)	rec. GH 1.75mg 3.5mg 7.0mg (sc 1x/d)	25 50 100	12	63 (1 to 63)	Heifers 70kg 1.5mo	+7.1* +8.3* +10.8* (1.2)	+2.5 +2 +2.5	+3.8 +5.7* +7*	+1.2 +2.5* +0.8 (59.3)	N.D.	N.D.	N.D.	N.D.
Grings et al. (1987)	50mg rec.GH (im 1x/d)	170	40	154 (1 to 154)	Heifers 295kg 13-18mo	+25* (.68)	N.D.	N.D.	N.D.	N.D.	N.D.	N.D.	N.D.
Wagner et al. (1988, pers.comm)	960mg rec.GH (sc 1x/14d)	204	14	140 (1 to 140)	Steers 336kg	+10* (1.25)	-8*	+19*	-3* (63)	+1 (321)	+9*	-25*	N.D.
McShane et al. (1988)	25mg GH (sc 1x/d)	140	10	250 (150 to 250)	Heifers 7mo	+*	N.D.	N.D	N.D.	N.D.	N.D.	N.D.	N.D.
Hancock and Preston (1988)	GH 8mg 16mg 32mg (sc 1x/d)	20.5 41 82	6	21 (1 to 21)	Steers 389kg	+52 +295* +248* (.21)	N.D.	N.D.	N.D.	N.D.	N.D.	N.D.	N.D.
Early et al. (1988)	25mg rec.GH (sc 1x/d)	108	10	112 (1 to 112)	Steers 231kg	+15* (1.2)	None	+*	-4* (54)	None (201)	None	–	N.D.
Crooker et al. (1988)	rec. GH 0.7mg 3.4mg 7.0mg 10.4mg 20.8mg (im 1x/d)	6.7 33 67 100 200	6	13 (6 to 12)	Heifers 104kg	N.D.	N.D.	N.D.	N.D.	N.D.	N.D.	N.D.	+[f] + + + +23* (25)

*Where possible, actual control values for ADG(kg), KO % (%), carcass wt. (kg) and N ret. (g/d) appear in parentheses under the % change. BW = body weight; trt. = treatment; d = day; wt. = weight; ADG = average daily gain; ADFI = average daily feed intake; FCE = feed conversion efficiency; KO% = kill-out percent; N ret. = nitrogen retention; pit. = pituitary; rec. = recombinant; GH = growth hormone; sc = subcutaneous; im = intramuscular; iv = intravenous; x = time; N.D. = not determined or available; * = significant (P<.05); mo = month; cont. = continuous; inj. = injected; N.S. = not significant; R = restricted feed; AL = ad libitum feed.

[b] Used identical twin pairs.   • Quadratic dose response.
[c] Energy conversion efficiency.   [f] Significant curvilinear dose response.
[d] Eight week withdrawal before slaughter.

subsequent studies in Yugoslavia (Car et al., 1967; zNidar, 1976) using high doses of pituitary GH in older animals failed to detect a significant effect of GH on growth in steers and bulls. However, duration of treatment was short and number of animals per treatment small. In a study similar to that of Brumby (1959), Sejrsen et al. (1986, personal communication) observed a significant 8.6% increase in ADG in heifers treated with a highly purified GH preparation, which confirms the result of Brumby (1959).

In a relatively long duration experiment with growing steers Quirke et al. (1985, personal communication) noted a significant, albeit small (6%), increase in ADG due to daily treatment with a low dose of pituitary GH. Wagner et al. (1988, personal communication) in a similar study observed a significant 10% increase in ADG in response to GH. However, in this latter study the daily dose was approximately five-fold higher than that used by Quirke et al. (1985, pers. comm.) and was administered as a subcutaneous bolus injection every 14 days. Both studies noted an additive effect of GH with an oestradiol implant for ADG and feed efficiency (but not for carcass measurements). In a third long term steer study, Early et al. (1988) obtained a 15% increase in ADG in response to GH using a daily dose intermediary between that of the other two studies mentioned. Early et al. (1988) and Wagner et al. (1988, pers. comm.) used recombinant GH in their studies.

In a long term study with young dairy heifers, Sandles and Peel (1987) observed a significant 8.6% growth advantage in heifers treated with a very high daily dose of pituitary GH. In contrast, increases in ADG of 23-25% were achieved in growing (Grings et al., 1987) and finishing heifers (Fabry et al., 1987) using much smaller daily doses of GH. In the longest (250 day) GH growth trial to date in cattle, McShane et al. (1988) noted a significant improvement in ADG of heifers using a moderate daily dose of GH.

Williams et al. (1987) found no effect of GH in very young bull calves during a 100 day treatment period. In contrast, using young heifer veal calves, Kirchgessner et al. (1987) obtained significant improvements in gain (8.7%)

with three different doses of recombinant GH during a 63 day trial. The doses used, which did not differ in effect, were similar to that used by Williams et al. (1987). Other dose response studies have been of relatively short duration. For example, Wolfrom and Ivy (1985) conducted two dose response studies over an 8 day period using 9-12 growing steers per treatment. In the first study ADG was significantly increased similarly by 50 and 100 ug GH/kg BW/day. In contrast, in the second study 19,38 and 56 ug GH/kg BW/day were without effect on ADG. This demonstrates the variable response between studies which can be achieved using nearly identical materials and methods. A third study by the same group (Ivy et al., 1986) using nearly identical doses and similar animals as in the second study, but lasting for 42 days, demonstrated a significant increase (quadratic dose response) in ADG in response to GH. In a similar (i.e., doses and steers) short term (21 day) study with only six animals per treatment Hancock and Preston (1988) reported large increases in growth rate.

One study has examined the effect of GH treatment in restricted fed (1.3% of BW; DM) versus ad libitum fed animals (Peters, 1986). This study employed a medium dose of GH administered to growing steers for 29 days. Restricted fed steers increased ADG 20% while ad libitum fed steers decreased ADG 6% in response to GH treatment.

Two studies investigated growth rate after cessation of GH treatment. Sandles and Peel (1987) found that BW of treated and control heifer groups were similar by five weeks post-treatment. In contrast, Quirke et al. (1985, pers. comm.) observed that the BW advantage of the GH-treated steers was maintained for at least eight weeks.

In summary, in longer term growth trials, GH treatment increases growth rate approximately 12% (range 3-25%). Although the response varies considerably between studies, GH consistently has a positive effect on growth of cattle.

Feed intake and efficiency

Unfortunately, feed intake has not been measured in all the growth trials and in one study (Sejrsen et al., 1986, pers. comm.) animals were pairfed. (Table 1). In all but one study in which feed intake was ad libitum and recorded it was not significantly affected by GH treatment. The exception to this is the study of Wagner et al. (1988, pers. comm.) who observed a significant 8% decrease in intake throughout the 140 day trial.

Because the effects of GH on intake are generally small or absent, the magnitude of improvement in feed (or energy/protein) efficiency noted tends to parallel the effects on ADG. However, depending on the study, the effect of GH on feed efficiency may actually be less (Sandles and Peel, 1987), the same (Fabry et al., 1987) or greater (Wagner et al., 1988, pers. comm.) than the effect on growth.

Carcass characteristics

Only eight studies have examined the effects of GH on carcass weight and/or kill-out percent (KO%) (Table 1). Generally, GH had either no effect or induced a small non-significant increase in carcass weight. The one exception is Sandles and Peel (1987) who obtained a significant 9% increase in carcass weight due to GH treatment. They slaughtered only four heifers per treatment; however, they did use identical twin pairs which would improve the confidence in the result found. Interestingly, this study employed the largest dose used in any trial, a long duration of treatment and young animals suggesting that the increased carcass weight may have been due to an effect of GH on skeletal development in the growing animal. However, GH did not significantly alter wither height or hip width, although chest girth was increased. It is important to note that if effects on carcass weight were expressed as average daily carcass gain, the percentage increase due to GH would be larger and likely significant in several studies. Most studies have found no effect of GH on KO%. However, one study (Kirchgessner et al., 1987) found a significant increase while two studies (Wagner et al., 1988, pers. comm.; Early et al., 1988) found significant decreases in KO% as a result of GH treatment.

Six studies have measured some indicator of carcass composition (Table 1). The proportion of lean in the carcass is either not affected (Sandles and Peel, 1987; Fabry et al., 1987; Early et al., 1988) or increased (Sejrsen et al., 1986, pers. comm.; Peters, 1986; Wagner et al., 1988, pers. comm.) by GH administration. The difference in responses can not obviously be attributed to type or dose of GH, duration of treatment or type of animal. Importantly, if growth rate is increased coincident with either no effect or an increase in percent carcass lean then the yield of carcass lean will be increased. The proportion of fat in the carcass is generally significantly decreased by GH treatment, by as much as 32%. In five of the six studies ADG was significantly increased and this may or may not be accompanied by an increase in lean or a decrease in fat. For example, Fabry et al. (1987) observed no effect of GH on carcass lean or fat, even though they used finishing heifers and observed an increase in growth rate of 23%.

Nitrogen retention

Six studies (most short term) have examined the effect of GH on nitrogen (N) retention (Table 1). Generally, the effect was positive and significant with increases ranging from 2 to 500%. In the study of Crooker et al. (1988) N retention was linearly increased using increasing doses of GH. The large variation in the magnitude of the percentage increase in N retention is likely a function of the actual N retention of control animals. Both carcass lean and N retention were measured in only one study (Sandles and Peel, 1987). They found no effect of GH either on carcass lean or N retention measured towards the end of the 147 day trial.

Puberty and reproduction

Two recent abstracts have addressed the effect of GH on puberty and reproduction in growing heifers. Mc Shane et al. (1988) found no effect of daily GH treatment for 250 days on puberty in a situation where ADG was significantly increased. However, the area of the pelvic opening was increased by GH. Similarly, Grings et al. (1987) observed a 23% increase in pelvic opening size which coincided with a 19% increase in hip height (and a 25% increase in ADG) when heifers were treated daily with GH for 154 days. In addition, Grings et al. (1987) found no effect of GH on conception rate when heifers were bred 60 days after commencement of GH treatment.

SHEEP
Growth rate

Wheatley et al. (1966) was the first study to document the effect of GH on growth in sheep (Table 2). They reported no effect of GH on growth during a 28 day trial. This result was not surprising because they used ten year old ewes and growth effect was not a primary objective of their study. However, two later studies where the effect of GH on wool growth was the primary objective also found no effect on growth (Reklewska, 1974). Although very young lambs were used, the duration of treatment was 90 - 100 days and the daily dose of bovine GH seemed adequate, the lack of effect may have been due to the fact that lambs were injected with GH only every 10 - 15 days.

Wagner and Veenhuizen (1978) were the first to report a positive significant effect of GH on growth in sheep. They used heavy wether lambs and probably optimized their chance of success by using a high daily dose of ovine GH given twice daily for 98 - 112 days. Subsequently, Muir et al. (1983) used a larger number of lambs and reported only a small non-significant effect of ovine GH on growth. The major methodological differences between these two studies is that Muir et al. (1983) used a somewhat smaller daily dose of GH, injected lambs only once daily, used olive oil as the vehicle, used lighter lambs and went for 56 days. Surprisingly perhaps, blood concentrations of GH in lambs were twice as high (due to GH treatment) in Muir et al. (1983) than in Wagner and Veenhuizen (1978).

Wolfrom et al. (1985) treated lambs with three daily doses of GH for 28 days and reported a linear dose - related increase in growth up to a maximum of 22% which was significant above controls. Three studies conducted by Johnsson and co-workers in the U.K. have yielded conflicting

TABLE 2 The effects of exogenous growth hormone on growth and carcass characteristics of sheep[a].

Reference	Treatment	Daily dose (ug/kg BW)	No. per trt.	Duration of trt., d (period analysed)	Anim type, initial wt./age	ADG	ADFI	FCE	KO%	Carcass wt.	Carcass Lean	Carcass Fat	N ret.
Struempler and Burroughs (1959)	12.5mg pit.bGH (im 1x/d)	357	72	12 (1 to 12)	Wether lambs 35kg	N.D.	N.D.	N.D.	N.D.	N.D.	N.D.	N.D.	+50* (3.1)
Wallace and Bassett (1966)	10mg pit.oGH[b] (sc 1x/d)	200	4	28 (1 to 28)	Ewes 50kg	N.D.	N.D.	N.D.	N.D.	N.D.	N.D.	N.D.	+*
Wheatley et al. (1966)	5mg pit.GH (sc 1x/d)	167	4	28 (1 to 28)	Ewes 30kg 10yr	None	N.D.	N.D.	N.D.	N.D.	N.D.	N.D.	+44 (.4)
Davis et al. (1970)	5mg pit. oGH (iv 2x/d)	286	4	9 (1 to 9)	Wether lambs 35kg	N.D.	N.D.	N.D.	N.D.	N.D.	+*	-*	+159* (2.5)
Vezinhet (1973)	pit.bGH (3x/7d)	1286	14	75 or 100 (1 to 75/100)	Ewe/ram lambs 25d	N.D.	N.D.	N.D.	N.D.	N.D.	N.D.	N.D.	N.D.
Reklewska (1974)	1mg/kg per 15d pit.bGH per 10d	67 100	9 8	90 100	Ewe lambs 10d	None None	+35* +29*	N.D.	N.D.	N.D.	N.D.	N.D.	N.D.
Wagner and Veenhuizen (1978)	7.5mg pit.oGH (s.c. 2x/d)	375	5	98 or 112 (1 to 98/112)	Wether lambs 40kg	+20* (.19)	None	+14*	N.D.	N.D.	+25*	-36*	N.D.
Muir et al. (1983)	7mg pit.oGH (sc 1x/d in oil)	250	16	56 (1 to 56)	Wether lambs 28kg	+4 (.27)	-4	+7.4*	N.D.	+1.5 (26)	+6.5	-9*	N.D.

TABLE 2 (continued).

Reference	Treatment	Dose	N	Days (range)	Animal	Col1	Col2	Col3	Col4	Col5	Col6	Col7	Col8
Johnsson et al. (1985)	1.7mg pit.bGH (sc 1x/d)	100	8	84 (1 to 84)	Ewe lambs 17kg 8wk	+22* (.28)	+6	+14*	-5 (52)	+6 (22)	+24*	-12	N.D.
Wolfrom et al. (1985)	pit.bGH 1mg/lamb (inj. 5mg/lamb 1x/d) 10mg/lamb	N.D.	8	28 (1 to 28)	Lambs	+c + +22*	None None None	+c + +15	N.D.	N.D.	N.D.	N.D.	N.D.
Pullar et al. (1986)	0.1mg/kg rec.bGH (sc 1x/d)	100	4	42 (1 to 42)	Ram lambs	+30* (.24)	N.D.	N.D.	N.D.	+12 (16)	+1.4	-2	N.D.
Johnsson et al. (1987)	rec.bGH .5mg (sc 1x/d) 2mg 5mg	25 100 250	8-10	84 (1 to 84)	Ewe/wether lambs 20kg 10wk	+3.1 -4.4 +1.3 (.23)	0 -3.9 -5.7	+3 0 +7.2	0 -3.2 -9.3 (53) (N=4)	+3.1 -4.6 -7.2* (19.5) (N=4)	-9.4 +1 +5 (N=4)	+10 -3.8 -14.5	N.D. N.D. N.D.
Pell and Bates (1987)	2mg rec.bGH (sc 1x/d)	100	10	84 (1 to 84)	Ewe/wether lambs 9wk	+38* (.23)	N.D.	N.D.	N.D.	N.D.	+6	N.D.	N.D.
Beerman et al. (1988)	1mg oGH (sc 4x/d)	160	19	42 or 56	Lambs 25.5kg	N.D.	N.D.	N.D.	N.D.	None	+8*	-19*	N.D.
Heird et al. (1988)	2.5mg oGH (sc 1x/2d)	N.D.	15	98 (1 to 50) (51 to 98)	Ewe lambs 4mo	+* None	None None	+* None	N.D.	N.D.	N.D.	N.D.	N.D.
Wise et al. (1988)	2.5mg oGH (sc 2x/d)	200	12	42 (1 to 42)	Wether lambs 25kg	+12* (.42)	N.D.	+20*	-*	None	+*	-*	N.D.

a See footnote at the bottom of Table 1.
b Ovine GH.
c Significant linear dose response.

results. In the first of these studies (Johnsson et al., 1985), growing ewe lambs treated with a moderate daily dose of pituitary bovine GH for 84 days had 22% better ADG than controls. Similarly, in a later study (Pullar et al., 1986), slightly older ram lambs treated with approximately the same daily dose of recombinant bovine GH for 42 days increased ADG 30% above controls. However, in a third similar study (Johnsson et al., 1987), the same group observed no effect of GH on lamb growth in spite of the fact that three doses of GH were tested (doses which were similar to those used successfully by Wolfrom et al., 1985). Furthermore, using the same design and materials as Johnsson et al. (1987), Pell and Bates (1987) reported a 38% increase in ADG. The reason for these discrepancies is unknown.

Two reports have recently been presented to the 1988 American Society of Animal Science meeting which addressed the effect of ovine GH on growth of lambs. Wise et al. (1988) achieved improvement in gain of 12% during the course of a 42 day trial where a high dose of GH was injected twice daily. Similarly, Heird et al. (1988) also demonstrated a significant increase in ADG of lambs during a 50 day treatment period, although they utilised a low daily dose of GH administered only every second day. Interestingly, when the experiment was continued for another 48 days no effect of GH on growth was noted. This may have been due to a dilution effect of exogenous GH in the body as dose of GH was not adjusted upwards over time.

In summary, in longer term growth trials with growing lambs, GH treatment increases growth rate approximately 18% (range 0 - 38%). Although the response varies considerably between studies, GH usually has a positive effect on growth of sheep.

Feed intake and efficiency

In most of the growth trials feed intake and efficiency have been measured (Table 2). In all but one report, feed intake was not significantly affected by GH treatment. The exception to this are the studies of Reklewska (1974) who observed significant increases in feed intake of approximately 33 % during GH treatment even though growth rate was unaffected. Although not reported, these data imply that FCE was depressed in GH-treated lambs. Of interest, but unexplainable, is the additional observation that feed intake of GH treated lambs was still 30% higher than controls 15 weeks after cessation of GH treatment.

Because the effects of GH on intake are generally small or absent, the magnitude of improvement in feed efficiency noted tends to parallel the effects on ADG. However, depending on the study, the effect of GH on feed efficiency may actually be less (Johnsson et al., 1985), the same (Johnsson et al., 1987) or greater (Muir et al., 1983) than the effect on growth.

Carcass characteristics

Only six studies have examined the effects of GH on carcass weight and/or KO% (Table 2). Generally, GH has little or no effect on carcass weight. The only significant effect noted was by Johnsson et al. (1987) who found a dose - related decrease in carcass weight due to GH when growing lambs were treated for 84 days. Growth of lambs in this study was not altered by GH and thus KO% was also decreased in a dose - dependent manner. Pullar et al. (1986) reported a 12% increase, albeit non-significant, in the carcass weight of ram lambs which had increased ADG of 30% due to GH treatment. Only three studies reported on the effect of GH on KO%. One was the study of Johnsson et al. (1987). The other two (Johnsson et al., 1985; Wise et al., 1988) also reported a decrease in KO%. The small but frequent decrease in KO% in sheep and cattle is likely due to a decrease in abdominal fat.

In nine studies some indicator of carcass composition has been measured (Table 2). The proportion of lean in the carcass is either not affected (Muir et al., 1983; Pullar et al., 1986; Johnsson et al., 1987; Pell and Bates, 1987) or increased (Vezinhet, 1973; Wagner and Veenhuizen, 1978; Johnsson et al., 1985; Beerman et al., 1988; Wise et al., 1988) by GH administration. The differences in response can not obviously be attributed to dose of GH, duration of treatment or type of animal. However, interestingly, three of the four studies in which no effect of GH on carcass lean was observed (Pullar et at., 1986; Johnsson et al., 1987; Pell and Bates, 1987) used recombinant bovine GH. To my knowledge, no direct comparisons between pituitary and recombinant GH or between ovine and bovine GH have been made in growth and carcass composition studies in sheep. Johnsson et al. (1987) noted that carcass lean content improved (non-significantly) as the dose of GH increased. The proportion of fat in the carcass was either not affected (Johnsson et al., 1985; Pullar et al., 1986; Johnsson et al., 1987) or decreased (Vezinhet, 1973; Wagner and Veenhuizen, 1978; Muir et al., 1983; Beerman et al., 1988; Wise et al., 1988) by GH administration. In five of the nine studies ADG was significantly increased and this may or may not be accompanied by an increase in lean or a decrease in fat. For example, Pullar et al. (1986) observed no effect of GH on carcass lean or fat, even though GH increased ADG 30%. Possibly the use of ram lambs and the duration of treatment were significant in this respect.

Nitrogen Retention

Four short term (9 - 28 days) studies have examined the effect of GH on N retention (Table 2). In all studies increases due to GH were observed and only one (Wheatley et al., 1966) was not significant. Wheatley et al. (1966) may not have seen a significant increase due to the fact that they utilized ten year old ewes.

Wool Growth

In nearly all studies in which wool growth has been monitored a significant effect of GH on fleece weight has been observed (Ferguson, 1954; Wheatley et al., 1966; Reklewska, 1974; Wynn, 1982; Johnsson et al., 1985; 1987). The exception is Muir et al. (1983) who saw no effect on wool growth of lambs during treatment with a high dose of ovine GH given daily for 56 days. Of the studies which have shown a significant effect, some observed an increase during treatment (Ferguson, 1954; Johnsson et al., 1985; 1987), some observed a decrease during treatment (Wheatley et al., 1966; Wynn, 1982) and some observed an increase after cessation of GH treatment (Ferguson, 1954; Wheatley et al., 1966; Reklewska, 1974; Wynn 1982). Interesting is the observation that when wool growth was depressed during treatment and increased post-treatment ovine GH was used, whereas bovine GH was used in studies where wool growth was increased during the treatment period (Hart and Johnsson, 1986).

GROWTH HORMONE - RELEASING FACTOR

Table 3 provides a summary of materials, methods and results for all the studies in cattle and sheep which have examined the effects of exogenously administered GRF on growth performance and carcass characteristics.

Only two experiments in cattle have examined the effect of GRF on growth. One study (Enright, unpublished) utilized finishing heifers which were treated daily with a GRF analogue (approximately ten times the potency of human GRF 1 - 44 in terms of GH release) for 84 days. Small increases in growth, KO% and carcass weight due to GRF were observed. In the long term trial of Ringuet et al. (1988) with growing heifers there was no effect on growth but GRF affected carcass composition of cattle in a manner similar to GH. Three studies have examined the effect of GRF on N retention in cattle. In all three studies there were increases due to GRF but in only one study (Moseley et al., 1987) was the increase significant.

In two studies the effect of GRF on growth of sheep was examined. Pastoureau et al. (1988) found no effect in very young ram lambs treated for 90 days. In contrast, Wise et al. (1988) observed significant increases in ADG and FCE due to GRF where lambs were injected two or four times a day. They also found a significant decrease in KO% due to GRF. Two similar studies (Wise et al., 1988; Beermann et al., 1988), where a total of 62 lambs received GRF, clearly demonstrated that GRF had no effect on carcass weight, significantly increased carcass lean and significantly

TABLE 3 Effects of growth hormone - releasing factor on growth and carcass characteristics of cattle and sheep.[a]

Reference	Treatment	Daily dose (ug/kg BW)	No. per trt.	Duration of trt.;d (period analysed)	Anim type, initial wt./age	ADG	ADFI	FCE	KO%	Carcass Wt.	Carcass Lean	Carcass Fat	N ret.
Moseley et al. (1987)	3.6mg hGRF (iv cont.)	24.3	7	20 (9 to 14)	Bulls 148kg	N.D.	R.	N.D.	N.D.	N.D.	N.D.	N.D.	+18* (41)
Plouzek et al. (1988)	6ug rGRF (iv 6x/d)	0.4	5	10 (1 to 10)	Bulls 90kg 3.5mo	N.D.	N.D.	N.D.	N.D.	N.D.	N.D.	N.D.	+16 (16)
Ringuet et al. (1988)	5ug/kg hGRF (sc 2x/d)	10	12	245 (1 to 245)	Heifers 3mo	None	N.D.	N.D.	N.D.	None	+9*	-7*	N.D.
Lapierre et al. (1988)	730ug hGRF (sc 2x/d)	10	10	73 or 82 (last 7)	Bulls 146kg	N.D.	N.D.	N.D.	N.D.	N.D.	N.D.	N.D.	+35 (31)
Enright, unpublished	343ug hGRF[b] (sc 1x/d)	1	10	84 (1 to 84)	Heifers 343kg	+7 (1.2)	N.D.	N.D.	+1 (51)	+3 (221)	N.D.	N.D.	N.D.
Pastoureau et al. (1988)	34.4ug GRF (sc 2x/d)	16	6	45(1 to 45) 90(1 to 90)	Ram lambs 4.3kg	None None	N.D.	N.D.	N.D.	N.D.	N.D.	-* None	N.D.
Beerman et al. (1988)	hGRF 127.5ug (sc 255ug 4x/d)	20 40	19	42 or 56	Lambs 25.5kg	N.D.	N.D.	N.D.	N.D.	None None	+6* +10*	-14* -20*	N.D.
Wise et al. (1988)	hGRF.5mg 2x/d .25mg4x/d	40	12	42 (1 to 42)	Wether lambs 25kg	+12* +18*	None None	+12* +18*	-* -*	None None	+* +*	-* -*	N.D.

[a] See footnote at the bottom of Table 1.
[b] [desamino-Tyr¹, D-Ala², Ala¹⁵] hGRF1-29-NH₂

decreased carcass fat.

Overall, from the available evidence in cattle and sheep, GRF affects growth and carcass characteristics in a manner similar to GH and thus, because of its smaller size, GRF may prove to be an even more attractive commercial growth/carcass enhancer than GH.

NUTRIENT REQUIREMENTS

In growing cattle administration of GH does not affect nutrient digestibility or total heat production but rather has a dramatic effect, direct or indirect, on the partitioning of absorbed nutrients (Bauman and McCutcheon, 1986; Eisemann et al., 1986b). In cattle in positive energy balance receiving exogenous GH, the shift in nutrient partitioning appears to involve a decrease in energy deposition in tissue to accommodate increased nutrient utilization for lean tissue accretion (Bauman and McCutcheon, 1986). Similarly, in growing cattle fed at maintenance intake GH administration partitions nutrients to protein accretion at the expense of increased turnover of non-esterified fatty acids (NEFA) in adipose tissue (Eisemann et al., 1986b).

The detailed experiment of Peters (1986) with growing steers is the only study to examine the effect of GH in animals on two different planes of nutrition (Table 1). In steers fed ad libitum, GH over a 29 day period decreased ADG slightly (6%), decreased feed intake 14% (not significant) and had large effects on carcass composition (i.e., increased lean, decreased fat). In contrast, GH treatment of restricted fed steers increased ADG 20% (not significant) but had only a small effect on carcass composition. Thus, plane of nutrition may have a big effect on response attained by GH administration. Peters (1986) concluded that sufficient nutrients must be present for GH to effect a net decrease in lipid accretion. If dietary energy is limited, GH acts (directly or indirectly) on adipose tissue to provide the necessary energy as NEFA. Importantly, irrespective of feeding level, GH can promote protein accretion to some extent. Because of the obvious effect of

GH on increasing body nitrogen retention and carcass lean content, an adequate level of protein intake will be critical to achieve optimum results with GH. To date, no studies have addressed this issue in growing ruminants.

Recent work by Gluckman and co-workers (Breier et al., 1988a, 1988b) demonstrated that adequate nutrition is necessary to achieve optimal use of GH. Relative to maintenance - fed steers, well - fed steers have higher basal and GH - induced concentrations of somatomedin - C (insulin - like growth factor 1) and affinity of hepatic GH receptors, both of which are considered to be very important for full biological participation of GH in the growth process.

In respect to mineral requirement in the GH - treated ruminant, little work has been done but GH is thought to co-ordinate mineral partitioning (Bauman and McCutcheon, 1986). Braithwaite (1975) observed increased apparent absorption and net accretion of calcium and phosphorus in GH-treated lambs. Eisemann et al. (1986a) found no effect of GH on serum concentrations of calcium in heifers, although serum phosphorus was reduced but remained well within the normal range. In a long-term trial in lactating cows several doses of GH had no effect on body mineral content or on concentrations of calcium, phosphorus, sodium, chloride and potassium in blood. To my knowledge, no study has addressed the issue of vitamin requirement for the GH-treated animal.

In summary, much work still needs to be done to examine the nutrient requirements of ruminants treated with GH. However, from the available evidence, it appears that good nutrition (and especially adequate protein) will be necessary to achieve optimal growth and carcass responses to GH administration.

SUMMARY AND CONCLUSIONS

This paper has attempted to review, present and discuss the available data on the effects of GH and GRF on growth, efficiency and carcass composition of ruminants. It was felt necessary to include data from abstracts and personal communications. To my knowledge 9 papers/11 abstracts and 10 papers/6 abstracts exist for cattle and sheep, respectively, on the effect of exogenous GH on growth and (or) nitrogen retention. The respective numbers for GRF are 2 papers/2 abstracts/1 personal communication and 1 paper/2 abstracts. No pertinent data for the goat has been found. From the reference lists in the tables it is clear that this area of research is accelerating due to several factors such as the advent of recombinant DNA technology, the plethora of recent studies in the lactating cow and the need for an alternative to steroid implants in the beef industry (especially in the EC countries) to improve efficiency and decrease carcass fat.

Based on the data reviewed in this paper it appears that administration of GH to cattle will increase ADG 12%, FCE 9% and carcass lean content 5%, while decreasing carcass fat content 15%. Similarly, administration of GH to sheep will increase ADG 18%, FCE 14% and carcass lean content 10%, while decreasing carcass fat content 15%. Thus, GH appears to satisfy the commercial requirements of an effective growth/carcass enhancer for ruminants. However, an insufficient number of long term studies have been published to date. There is huge variation in response to GH between studies, likely due to the fact that studies are very diverse in design, materials and methodology used. Many important basic and applied experiments still need to be performed to examine the effect of potential sources of variation on response attained. Sources of variation that need to be addressed include source of GH (pituitary versus recombinant; ovine versus bovine); dose and biological potency of GH preparations; route and frequency of administration; sex, gonadal status, type and age of animals; duration of treatment; level and type of nutrition (especially dietary protein supply); and length of withdrawal period before slaughter. Other studies needed include the response to GH in the presence of other nutritional and physiological manipulations such as ionophores and immunization against somatostatin.

ACKNOWLEDGEMENTS

I thank Mary Smith and Mary O'Brien for their help in preparing the manuscript. I also thank Cathryn Geraghty, Charlie Godson and John Dowling for their assistance in preparing the slides used in the presentation.

REFERENCES

Anonymous. 1984. Diet and cardiovascular disease. Department of Health and Social Security. Report on Health and Social Aspects, No. 28. H.M.S.O., London.

Bauman, D.E., and McCutcheon, S.N. 1986. The effects of growth hormone and prolactin on metabolism. In "Control of digestion and metabolism in ruminants". (Eds. L. P. Milligan, W. L. Grovum and A. Dobson). Proceedings of the Sixth International Symposium on Ruminant Physiology. Prentice-Hall, New Jersey. pp. 436-455.

Beerman, D.H., Hogue, D.E., Fishell, V.K., Dickson, H.W., Aronica, S., Dwyer, D. and Schricker, B.R. 1988. Effects of exogenous ovine growth hormone ($_o$GH) and human GH releasing factor (hGRF) on plasma $_o$GH concentration and composition of gain in lambs. J. Anim. Sci. 66 (Suppl. 1) : 282-283 (Abstr.).

Braithwaite, G. D. 1975. The effect of growth hormone on calcium metabolism in the sheep. Br. J. Nutr. 33 : 309-314.

Breier, B. H., Gluckman, P. D. and Bass, J. J. 1988a. Influence of nutritional status and oestradiol-17B on plasma growth hormone, insulin - like growth factors -I and -II and the response to exogenous growth hormone in young steers. J. Endocr. 118 : 243-250.

Breier, B. H., Gluckman, P. D. and Bass, J. J. 1988b. The somatotrophic axis in young steers : influence of nutritional status and oestradiol-17B on hepatic high- and low- affinity somatotrophic binding sites. J. Endocr. 116 : 169-177.

Brumby, P.J. 1959. The influence of growth hormone on growth in young cattle. N.Z. J. Agric. Res. 2 : 683-689.

Car, M., zNidar, A. and Filipan, T. 1967. [An effect of the treatment of young steers with STH (growth hormone) upon nitrogen retention in intensive feeding]. Veterinarski arhiv 5-6 : 173-184.

Crooker, B.A., Bauman, D.E., Cohick, W.S. and Harkins, M. 1988. Effect of dose of exogenous bovine somatotropin on nutrient utilization by growing dairy heifers. J. Anim. Sci. 66 (Suppl. 1) : 299 (Abstr.).

Davis, S.L., Garrigus, U.S. and Hinds, F.C. 1970. Metabolic effects of growth hormone and diethylstilbestrol in lambs. II. Effects of daily ovine growth hormone injections on plasma metabolites and nitrogen - retention in fed lambs. J. Anim. Sci. 30 : 236-240.

Early, R.J., McBride, B.W., Ball, R.O. and Rock, D.W. 1988. Growth, feed efficiency and carcass characteristics of beef steers treated with daily injections of recombinantly - derived bovine somatotropin. J. Anim. Sci. 66 (Supp. 1) : 283 (Abstr.).

Eisemann, J. H., Tyrell, H. F., Hammond, A. C., Reynolds P.J., Bauman, D.E., Haaland, G.L., McMurtry, J.P. and Varga, G.A. 1986a. Effect of bovine growth hormone administration on metabolism of growing Hereford heifers: dietary digestibility, energy and nitrogen balance. J. Nutr. 116 : 157-163.

Eisemann, J. H., Hammond, A. C., Bauman, D. E., Reynolds, P. J., McCutcheon, S. N., Tyrrell, H. F. and Haaland, G. L. 1986b. Effect of bovine growth hormone administration on metabolism of growing Hereford heifers : protein and lipid metabolism and plasma concentrations of metabolites and hormones. J. Nutr. 116 : 2504-2515.

Fabry, J., Claes, V. and Ruelle, L. 1987. Effect of growth hormone on heifer meat production. Reprod. Nutr. Develop. 27 : 591-600.

Ferguson, K. A. 1954. Prolonged stimulation of wool growth following injections of ox growth hormone. Nature, London. 174 : 411.

Grings, E.E., De Avila, D.M. and Reeves, J.J. 1987. Reproduction and growth in post-pubertal dairy heifers treated with recombinant somatotropin. J. Anim. Sci. 65 (Suppl. 1) : 248 (Abstr.).

Hancock, D.L. and Preston, R.L. 1988. Titration of bovine somatotropin (bST) dose response which maximizes plasma urea nitrogen (PUN) depression in feedlot steers. J. Anim. Sci. 66 (Suppl. 1) : 254 (Abstr.).

Hart, I.C. and Johnsson, I.D. 1986. Growth hormone and growth in meat producing animals. In " Control and manipulation of animal growth" (Eds. P.J. Buttery, D.B. Lindsay and N.B. Haynes). Butterworths, London. pp. 135-159.

Heird, C.E., Hallford, D.M., Spoon, R.A., Holcombe, D.W., Pope, T.C., Olivares, V.H. and Herring, M.A. 1988. Growth and hormone profiles in fine-wool ewe lambs after long-term treatment with ovine growth hormone. J. Anim. Sci. 66 (Suppl. 1) : 201 (Abstr.).

Ivy, R.E., Wolfrom, G.W. and Edwards, C.K. 1986. Effects of growth hormone and RALGRO (zeranol) in finishing beef cattle. J. Anim. Sci. 63 (Suppl. 1) : 217-218 (Abstr.).

Johnsson, I. D., Hart, I. C. and Butler - Hogg, B. W. 1985. The effects of exogenous bovine growth hormone and bromocriptine on growth, body development, fleece weight and plasma concentrations of growth hormone, insulin and prolactin in female lambs. Anim. Prod. 41 : 207-217.

Johnsson, I.D., Hathorn, D.J., Wilde R.M., Treacher, T.T. and Butler - Hogg, B.W. 1987. The effects of dose and method of administration of biosynthetic bovine somatotropin on live-weight gain, carcass composition and wool growth in young lambs. Anim. Prod. 44 : 405-414.

Kempster, A. J. 1988. Market requirements and their relation to carcass quality. In "Beta-agonists and their effects on animal growth and carcass quality". (Ed. J. P. Hanrahan). A Seminar in the CEC Programme of Coordination of Research in Animal Husbandry, Brussels, 1987. Elsevier Applied Science, London. pp. 72-82.

Kempster, A. J., Cook, G. L. and Grantley - Smith, M. 1986. National estimates of the body composition of British cattle, sheep and pigs with special reference to trends in fatness. A review. Meat Sci. 17 : 107-138.

Kirchgessner, M., Roth, F.X., Schams, D. and Karg, H. 1987. Influence of exogenous growth hormone (GH) on performance and plasma GH concentrations of female veal calves. J. Anim. Physiol. Anim. Nutr. 58 : 50-59.

Lamming, G. E. and others. 1987. Scientific report on anabolic agents in animal production. EEC Scientific Working Group on Anabolic Agents in Animal Production. Vet. Rec. 120 : 389-392.

Lapierre, H., Petitclerc, D., Pelletier, G., Dubreuil, P., Gaudreau, P., Couture, Y., Morisset, J. and Brazeau, P. 1988. Effects of growth hormone - releasing factor (GRF) and (or) thyrotropin - releasing factor (TRF) on energy and nutrient digestibilities and balances in male dairy calves. J. Anim. Sci. 66 (Suppl. 1) : 438 (Abstr.).

Moseley, W.M., Krabill, L.F. and Olsen, R.F. 1982. Effect of bovine growth hormone administered in various patterns on nitrogen metabolism in the Holstein steer. J. Anim. Sci. 55 : 1062 - 1070.

Moseley, W.M., Huisman, J. and Van Weerden, E.J. 1987. Serum growth hormone and nitrogen metabolism responses in young bull calves infused with growth hormone - releasing factor for 20 days. Dom. Anim. Endo. 4 : 51-59.

Muir, L.A., Wien, S., Duquette, P.F., Rickes, E.L. and Cordes, E.H. 1983. Effects of exogenous growth hormone and diethylstilbestrol on growth and carcass composition of growing lambs. J. Anim. Sci. 56 : 1315-1323.

McShane, T.M., Schillo, K.K., Boling, J.A., Bradley, N.W. and Hall, J.B. 1988. Effects of somatotropin and dietary energy on development of beef heifers. I. Growth and puberty. J. Anim. Sci. 66 (Suppl. 1) : 252-253 (Abstr.).

Pastoureau, P., Barenton, B., Blanchard, M., Boivin, G., Charrier, J., Dulor, J.-P. and Theriez, M. 1988. [Effects of GRF 1-29 in normal and hypotrophic lambs]. Reprod. Nutr. Develop. 28 : 253-256.

Pell, J. M. and Bates, P. C. 1987. Collagen and non-collagen protein turnover in skeletal muscle of growth hormone-treated lambs. J. Endocr. 115 : R1-R4.

Peters, J.P. 1986. Consequences of accelerated gain and growth hormone administration for lipid metabolism in growing beef steers. J. Nutr. 116 : 2490-2503.

Plouzek, C.A., Vale, W., Rivier, J., Anderson, L.L. and Trenkle, A. 1988. Growth hormone - releasing factor on growth hormone secretion in prepubertal calves. Proc. Soc. Exp. Biol. Med. 188 : 198-205.

Pullar, R.A., Johnsson, I.D. and Chadwick, P.M.C. 1986. Recombinant bovine somatotropin is growth promoting and lipolytic in fattening lambs. Anim. Prod. 42 : 433-434 (Abstr.).

Quirke, J.F., Kennedy, L.G., Roche, J.F., Hart, I. Sheehan, W., Coert, A. and Allen, P. 1985. Responses of finishing steers to exogenous growth hormone and oestradiol. Ir. Grassland and Anim. Prod. Assoc., 11th Ann. Res. Mtg.

Reklewska, B. 1974. A note on the effect of bovine somatotrophic hormone on wool production in growing lambs. Anim. Prod. 19 : 253-255.

Ringuet, H., Petitclerc, D., Sorenson, M., Gaudreau, P., Pelletier, G., Morisset, J., Couture, Y. and Brazeau, P. 1988. Effects of human somatocrinin (1-29) NH$_2$ (GRF) and photoperiod on carcass parameters and mammary growth of dairy heifers. J. Dairy Sci. 71 (Suppl. 1) : 193 (Abstr.).

Roche, J. F. and Quirke, J. F. 1986. The effects of steroid hormones and xenobiotics on growth of farm animals. In "Control and manipulation of animal growth" (Eds. P. J. Buttery, D. B. Lindsay and N. B. Haynes). Butterworths, London. pp. 39-51.

Sandles, L.D. and Peel, C.J. 1987. Growth and carcass composition of pre-pubertal dairy heifers treated with bovine growth hormone. Anim. Prod. 44 : 21-27.

Sejrsen, K., Foldager, J., Klastrup, S. and Bauman, D.E. 1986. Effect and mode of action of exogenous growth hormone on body tissue growth in heifers. 37th Ann. Mtg. EAAP 1: 467 (Abstr.).

Struempler, A.W. and Burroughs, W. 1959. Stilbestrol feeding and growth hormone stimulation in immature ruminants. J. Anim. Sci. 18 : 427-436.

Van Es. A.J.H. 1977. The energetics of fat deposition during growth. Nutr. Metab. 21 : 88-104.

Vezinhet, A. 1973. [Effect of hypophysectomy and bovine somatotrophic hormone treatment on the relative growth of lambs]. Ann. Biol. Anim. Bioch. Biophys. 13 : 51-73.

Wagner, J.F. and Veenhuizen, E.L. 1978. Growth performance, carcass deposition and plasma hormone levels in wether lambs when treated with growth hormone and thyroprotein. J. Anim. Sci. 45 (Suppl. 1) : 397 (Abstr.).

Wagner, J.F., Cain, T., Anderson, D.B., Johnson, P. and Mowrey, D. 1988. Effect of growth hormone (GH) and estradiol (E_2B) alone and in combination on beef steer growth performance, carcass and plasma constituents. J. Anim. Sci. 66 (Suppl. 1) : 283-284 (Abstr.).

Wallace, A. L. C. and Bassett, J. M. 1966. Effect of sheep growth hormone on plasma insulin concentration in sheep. Clin. Expt. Metab. 15 : 95-97.

Wheatley, I.S., Wallace, A.L.C. and Bassett, J.M. 1966. Metabolic effects of ovine growth hormone in sheep. J. Endocrin. 35 : 341-353.

Williams, P.E.V., Innes, G.M., Odgen, K. and James, S. 1987. The effects of a combination of the B agonist clenbuterol and bovine pituitary growth hormone on growth of milk-fed calves. Anim. Prod. 45 : 475 (Abstr.).

Wise, D.F., Kensinger, R.S., Harpster, H.W., Schricker, B.R. and Carbaugh, D.E. 1988. Growth performance and carcass merit of lambs treated with growth hormone releasing factor (GRF) or somatotropin (ST). J. Anim. Sci. 66 (Suppl. 1) : 275 (Abstr.).

Wolfrom, G.W. and Ivy, R.E. 1985. Effects of exogenous growth hormone in growing beef cattle. J. Anim. Sci. 61 (Suppl. 1) : 249-250 (Abstr.).

Wolfrom, G.W., Ivy, R.E. and Baldwin, C.D. 1985. Effects of growth hormone alone and in combination with RALGRO (Zeranol) in lambs. J. Anim. Sci. 61 (Suppl. 1) : 249 (Abstr.).

Wynn, P.C. 1982. Growth hormone and wool growth. Ph.D. Thesis, Univ. Sydney.

zNidar, A. 1976. [Effect of the application of growth hormone on the gains and feed consumption in simmental bulls]. Poljoprivredna znanstvena smotra 37 : 171-186.

...E OF SOMATOTROPIN IN LIVESTOCK PRODUCTION: GROWTH IN PIGS

T.J. Hanrahan

The Agriculture & Food Development Authority
Moorepark Research Centre, Fermoy, Co. Cork, Ireland

ABSTRACT

The ability of somatotropin to alter carbohydrate, fat and protein metabolism, resulting in a reduction in the amount of energy retained and an increase in the amount of lean tissue accretion, is of specific interest for meat production. Quality of pig meat is still strongly based on the percentage of lean meat and there is continuing emphasis in most countries to further reduce the percentage fat in the carcass.

Recombinant porcine somatotropin (rpST) has been shown, in a number of studies, to affect major improvements in growth performance and carcass composition of pigs. A dose level of 30 to 60 $ug.kgBW^{-1}.d^{-1}$ results in a c.5% reduction in feed intake, a c.20% improvement in growth rate and a c.20% reduction in the quantity of feed required per kg liveweight gain. The fat content of the carcass is reduced by about 30% and the lean content increased by about 10%.

Variation in the response to rpST recorded in different experiments and mobility problems recorded in two experiments indicate a need for further research to define more precisely the requirement for possible increased levels of amino acids, minerals and vitamins of treated animals.

Other factors affecting the response to rpST are genotype, sex, weight of pig and duration of treatment. These are discussed on the basis of the published information. Many of the references quoted in this review are abstracts presented at scientific meetings and refer to recent research. It is clear that a considerable amount of work is being undertaken on the subject and that within the near future the available literature will expand greatly.

INTRODUCTION

The initial discovery that secretions from the pituitary gland stimulated growth (Evans and Simpson, 1931) resulted in experimentation designed to elucidate the response rate and mechanisms involved. In one of the initial studies with pigs, Giles (1942) reported increased growth of bones on administering anterior pituitary extract. In 1955, Turman and Andrews recorded increased growth rate (16%), improved feed conversion efficiency (24%), decreased carcass fat (21%) and increased carcass protein (25%), when 11 pigs were treated with bovine pituitary extract. Henricson and Ullberg (1960) observed less dramatic results on administering porcine pituitary extract to pigs. In some of these experiments pigs exhibited severe antagonisms to the preparations administered. This was due to deficiencies in the procedures used for extracting and purifying pituitary extracts. As a consequence little progress was made.

Improvements in techniques during the 1960's and 70's resulted in the availability of more highly purified pituitary extracts and advances in protein chemistry allowed the elucidation of the molecular structure of growth hormone (somatotropin). The biochemistry and endocrinology of somatotropin has been reviewed by a number of authors (Daughaday, 1985; Laron, 1983). The breakthrough in genetic engineering of Cohen et al. (1973) has further developed to industrial production of recombinant somatotropin including recombinant porcine somatotropin. A number of commercial companies have produced recombinant somatotropin analogues which elicit the same response as pituitary porcine somatotropin. Animal production is, as a result, entering a phase of development that offers the possibility of more efficient production and greater control over the composition of meat and milk products.

Mode of action of somatotropin

The effect of somatotropin on growth is mediated, directly through alterations to carbohydrate, lipid and protein metabolism and indirectly by stimulating the production and secretion of somatomedin from the liver and other tissues. The altered carbohydrate metabolism in the liver results in an increase in blood glucose concentration and more glucose utilization in muscle. There is increased fat mobilization from adipose tissue to the liver and increased oxidation of fatty acids in liver and muscle. Intermediary metabolism of protein is altered, resulting in a net increase in protein tissue. The incorporation of more dietary amino acids into tissue protein results in a lowering of both blood urea concentration and urea excretion.

Somatomedin secreted mainly from the liver, binds with target tissue resulting in tissue cell proliferation (muscle, bone, connective and adipose tissue). It also stimulates an increase in protein synthesis. The increased heat production associated with protein synthesis results in a lowering of the percentage of dietary energy retained.

The ability of somatotropin to alter carbohydrate, fat and protein metabolism, resulting in a reduction in the amount of energy retained and an increase in the amount of lean tissue accretion, is of specific interest for meat production. Quality of pig meat is still strongly based on the percentage of lean meat and there is continuing emphasis, in most countries, to further reduce the percentage carcass fat. This is fueled by the movement of people from manual to sedentary occupations, with a requirement for lower energy foods to prevent obesity and other excess energy related conditions. Hence, the meat eating populations are demanding lower energy meats. Somatotropin adds a further tool to the existing array of factors such as genetics, sex, slaughter weight, feed scale and feed composition used to control the composition of meat.

Response to pituitary porcine somatotropin (ppST)

Following a series of experiments where daily injections of ppST were administered to pigs from 47 to 95 kg, Machlin (1972) reported that growth rate was improved by 12%, feed utilization efficiency was improved by 7 to 16%, backfat thickness was reduced by 2 to 15% and the percentage lean meat increased by 2 to 6%. The level of ppST used ranged from 33 to 132 ug.kg $BW^{-1}.d^{-1}$. The best response over all criteria was obtained at a level of 66 ug.kg $BW^{-1}.d^{-1}$. The quality of ppST used by Machlin was a cause of concern and he made reference to the variability in preparations, as measured by rat tibia assay, generally available at that time.

Chung et al. (1985) using a more potent preparation of ppST at a dose of 22 ug.kg $BW^{-1}.d^{-1}$ for a 30 day period (32 to 62 kg liveweight), reported no effect on feed intake and an improvement in growth rate of approximately 10%. The improvements in feed utilization efficiency and lean tissue growth rate were smaller and there was no reduction in carcass fat content. Rebhun et al. (1985), in a dose response experiment, obtained a 14% improvement in growth rate and a 17% reduction in feed required per kg gain, to the highest dose of 70 ug ppST.kg $BW^{-1}.d^{-1}$. This dose also resulted in a marked change in carcass composition with a reduction in carcass fat of 21% and an increase in muscle mass of 17%.

Pigs weighed 43 kg initially and were treated for 35 days.

Etherton et al. (1986), injected 30 ug ppST.kg $BW^{-1}.d^{-1}$ for a 30 day period (50 to 78 kg liveweight), and recorded a 10% reduction in feed intake, a 10% increase in growth rate and a 20% improvement in feed utilization efficiency. There was a marked change in carcass composition. Adipose tissue was reduced by 8% and muscle tissue increased by 36%. Administration of ppST resulted in an increase in blood glucose and insulin concentration and a reduction in blood urea concentration.

Etherton et al. (1987), in an experiment to determine dose response relationship, administered ppST at 10, 30 or 70 ug.kg $BW^{-1}.d^{-1}$ to pigs over the growth period 40 to 76 kg liveweight. The results indicated the maximally effective dose to be greater than 70 ug. At this dose feed intake was reduced by 5%, growth rate was increased by 14% and feed per kg gain reduced by 18%. Carcass quality was also significantly improved, with a 12% reduction in backfat thickness, a 13% increase in protein content and a 25% reduction in fat content of carcass tissue. The 70 ug dose of ppST also increased the weights of liver, heart and kidneys.

Boyd et al. (1986) in a similar experiment, but using ppST doses of up to 200 ug.kg $BW^{-1}.d^{-1}$, achieved the maximum growth rate response (+19%) at a dose of 60 ug. The feed requirement per kg gain was minimal (-28%) at a dose of 120 ug. Backfat thickness declined linearly, while eye muscle area and percentage muscle in the carcass increased linearly with increasing ppST dose. At the highest dose of ppST, the reduction in backfat was 33%, the increase in eye muscle area was 12% and the increase in muscle mass was 10%. McLaren et al. (1987), treated pigs over the growth period 57 to 103 kg liveweight, reported a similar dose response in that maximum growth rate was achieved at the 6 mg.pig^{-1}.day^{-1} level, whereas there was a linear improvement in feed utilization efficiency and carcass quality traits up to the highest level administered of 9 mg.pig^{-1}.d^{-1}. In contrast Wolfrom et al. (1986), in experiments with pigs treated from 50 to 85 kg liveweight, reported that the optimum dose was considerably lower at 2.3 to 2.7 mg.pig^{-1}.d^{-1}.

In two experiments using gilts (initial weight 72 kg), Bryan et al. (1987) reported faster growth rate and better feed efficiency and carcass quality on administering 70 ug ppST.kg BW^{-1}.d^{-1} for either a 31 or a 66 day period. However, he recorded locomotion and mobility problems, ranging from slight to severe, in 9 of 24 treated animals in the experiment. He also recorded significant increases in the weights of adrenal glands, liver and heart. A summary of published responses to ppST is shown in Table 1.

Response to recombinant porcine somatotropin (rpST)

Among the first to report on the use of rpST were Ivy et al. (1986), Kraft et al. (1986) and Etherton et al. (1986). All recorded responses which were similar to ppST in improving growth performance and carcass characteristics (Table 2). The experiments of Ivy et al. (1986) and Etherton et al. (1986) included direct comparisons between rpST and ppST. The former, treating pigs from 50 to 90 kg liveweight, recorded the maximum response in growth rate (+24%) at a dose of 4 mg.pig^{-1}.d^{-1}. Feed required per kg gain improved (-28%) up to the highest dose level of 8 mg.pig^{-1}.d^{-1}. Carcass quality was compared at the 0, 2 and 4 mg doses and there was a linear response to both rpST and ppST. Backfat thickness was

TABLE 1 Response(1) to pituitary porcine somatotropin (ppST)

Treatment as % of controls

Reference	Treatment	Pigs/treat.	Duration	Pig liveweight	A.D.F.I.(2)	A.D.G.(3)	F.C.E.(4)	KO%(5)	Backfat	LEA(6)
Rehbun et al. '85	10 to 70 ug ppST/kg BW/day	12	35 d.	43 kg initial		114	83		Reduced fat	Increased muscle
Chung et al '85	22 " " " " "	12	30 d.	32 to 62 kg	No effect	110	96	95	No effect	No effect
Etherton et al '86	30 " " " " "	12	30 d	50 to 78 kg	90	111	81		89	121
Etherton et al '87	10 to 70 " " "	12		40 to 76 kg	95	114	82		88	123
Boyd et al '86	30 to 200 " " "	8		48 kg initial	78	119	71		67	112
McLaren et al '87	1 to 9 mg ppST/pig/day	30		57 to 103 kg	77	114	70		42	113
Wolfrom et al '86	2 to 8 " " " " " "	5	35 d	50 kg initial	Reduced	No effect	Improved			

(1) Maximum recorded where increasing doses used
(2) Average daily feed intake
(3) Average daily gain
(4) Feed per kg gain
(5) Kill out %
(6) Loin eye area

TABLE 2   Response (1) to recombinant porcine somatotropin (rpST)

Treatment as % of controls

Reference	Treatment	Pigs/treat.	Duration	Pig liveweight	A.D.F.I.(2)	A.D.G.(3)	F.C.E.(4)	KO%(5)	Backfat	LEA(6)
Ivy et al '86	1 to 8 mg rpST/pig/day	12	42 d	50 kg initial	Reduced	124	72		78	127
Kraft et al '86	15 to 60 ug rpST/kg BW/day	24		55 to 105 kg		108	92.5		85	No effect
Etherton et al. '86	35 to 140 " " " " "	12	79 d	27 kg initial	Reduced	116	76		Reduced	145
Evock et al '88	35 to 140 " " " " "	12	77 d	27 to 110 kg	85	119	76	97.3	57	146

(1) Maximum recorded where increasing doses used
(2) Average daily feed intake
(3) Average daily gain
(4) Feed per kg gain
(5) Kill out %
(6) Loin eye area

reduced by 22% and loin eye area was increased by 27% at the highest dose. There were no adverse effects on health, even at the highest dose of rpST.

The work of Etherton et al. (1986) was reported in greater detail by Evock et al. (1988). Pigs were treated for a 77 day period from 27 to 107 kg liveweight. Treatments compared were 0, 35, 70 and 140 ug rpST and 35 and 70 ug ppST.kg BW^{-1} d^{-1}. Both forms of pST resulted in a linear reduction in feed intake, an improvement in growth rate (+19%) and an improvement in feed utilization efficiency (-20%) for the highest dose. Kill out % was reduced by about 2% as a result of increased weight of internal organs (and possibly blood volume) in pigs given 70 or 140 ug doses. The increase in muscle mass was similar (27%) for both preparations at the 70 ug dose, however ppST was more efficient at reducing carcass adipose tissue. For example at the 70 ug dose, the reduction in backfat was 49 and 18% for ppST and rpST respectively. The highest dose of 140 ug rpST resulted in impaired mobility of pigs, due to increased incidence of osteochondrosis. Interestingly there were no differences in calcium and phosphorous concentrations in serum or bones between the treatment groups. Some pigs on the 70 ug treatment groups also exhibited mobility problems.

We (Hanrahan and Allen, unpublished) have treated pigs (3 mg rpST.pig^{-1}.d^{-1}) for a 40 day period prior to slaughter at either 85 or 103 kg liveweight. Pigs were Large White x Landrace crosses and were either entire males or females. They were fed to appetite on a diet containing 3.2 Mcal DE/kg and 1% total lysine. Treated pigs consumed less feed (-5%), grew faster (+20%) and utilized their feed more efficiently (-20%). The liver, heart and kidney weights were higher in treated pigs by about 10 to 25%. Backfat thickness was reduced by about 20%. Carcass dissection indicated a 15% increase in skin, a 30% decrease in fat, an 8% increase in meat and a 4% increase in bone. rpST did not significantly affect meat pH measurements, chemical composition of muscle or muscle shear force. A summary of the main effects is shown in Table 3.

Factors affecting response to pST

While the response to pST is similar across all experiments there are minor differences, possibly due to slight variations in the potency of pST analogues, weight of pig, sex, genotype, nutrition and management of animals. Evock et al. (1988) refer to the fact that the rpST which they used was not entirely similar to ppST. This may explain the somewhat higher level 90 ug.kg $BW^{-1}.d^{-1}$ suggested to elicit the maximum growth response by these authors compared to 40 ug suggested by Ivy et al. (1986). Boyd et al. (1988) reported different biological activity between 2 variants of rpST.

Females showed a greater growth and carcass response than intact males to a dose of 100 ug ppST kg $BW^{-1}.d^{-1}$ (Campbell and Travener, 1988). There was a similar but not significant sex effect in the response observed by Hanrahan and Allen (unpublished) in that females showed the best carcass response.

In an unrelated but important aspect of the sex effect, Bryan et al. (1988) demonstrated that, while oestrus and ovarian activity are suppressed in ppST treated females withdrawal of treatment allows the expression of oestrus within 23 days. There appeared to be no effect on subsequent

TABLE 3 Performance of pigs treated with 3 mg rpST.pig^{-1}.d^{-1}[a]

Treatment period	48 to 84 kg			65 to 102 kg		
	Control	rpST	rpST as % Control	Control	rpST	rpST as % Control
Feed intake, day^{-1} kg	2.26	2.17	96	2.52	2.37	94
Gain, day^{-1} g	792	939	119	801	966	121
Feed/ kg gain kg	2.89	2.32	80	3.21	2.48	77
Kill out %	78.0	76.7	98	78.7	78.2	99
Heart + liver + lungs + kidneys kg	3.33	3.67	110	3.55	4.13	116
Skin % cold carcass	5.3	6.0	113	4.8	5.5	115
Meat % cold carcass	53.9	58.7	109	53.6	58.0	108
Fat % cold carcass	17.7	11.9	67	19.3	13.2	68
Bone % cold carcass	9.7	10.1	104	9.0	9.4	104
Head + feet + trim % cold carcass	11.9	12.2	103	11.8	12.1	103

[a]Hanrahan and Allen (unpublished)

ovulation or fertilization.

In general, heavier pigs appear to give a better response to pST than lighter pigs. Kanis et al. (1988) reported a better response in pigs slaughtered at 140 kg liveweight compared with 100 kg liveweight, while Evans et al. (1988) suggested that rpST has no effect on growth performance and little effect on carcass composition in pigs up to 50 kg liveweight.

The experiments of Huisman et al. (1988) and McLaughlin et al. (1988) indicate that genetically fatter pigs respond better. The former found little difference in the growth response between Pietrain, Duroc and Landrace x Yorkshire crosses, but the reduction in the fat content of the carcass was least for Pietrain. McLaughlin et al. (1988) reported that Beijing Black pigs showed greater responses to treatment, both for growth and carcass criteria, than U.S. type pigs.

Administration of pST

The duration of reported pST treatment varies from 30 to 79 days and there is no indication that the magnitude of the response alters greatly within this treatment period. Etherton et al. (1986) stated that somatotropin had a greater effect on growth performance and carcass composition when pigs were treated for a 79 day period by comparison with results, from previous experiments, where they treated pigs for a shorter period of 30 or 35 days. Bryan et al. (1987) recorded a better carcass response on administering 70 ug ppST.kgBW^{-1}.d^{-1} for a 66 rather than a 31 day period. Evock et al. (1988) suggest that the observed effect of rpST on pig mobility may be a factor of the high dose used and the length of the treatment period (77 days).

There is little in the literature on effect of pattern of administration on response to pST. Evans et al. (1988) reported on different administration patterns, upto 50 kg liveweight, but obtained no response to any of the treatments probably because of the age and weight of pigs used. Bryan et al. (1988) reported that withdrawal of treatment results in a subsequent loss of the growth performance advantage.

The successful use of an ear implant (6 week delivery life) has been reported by Knight et al. (1988). This development offers a practical administration system with minimum labour and handling stress for animals.

Nutrition of pST-treated animals

Because pST elicits such a major improvement in growth performance, the nutrition of treated animals obviously demands special attention. In particular, the protein-amino acid requirements are likely to be much higher because of increased protein retention. Huisman et al. (1988) recorded an increase in protein retention of 32% in treated animals. The maximum protein accretion achieved was 210 $g.pig^{-1}.d^{-1}$ in Dutch Landrace x Large White pigs fed an 18% crude protein diet.

Most experiments to date have been concerned with documenting the response to pST while using standard diets, or diets with additional lysine. An exception is Newcomb et al. (1988) who fed diets with protein levels ranging from 14 to 26%. The best growth performance was achieved with diets of 20 to 23% crude protein but the indications were that the lean content of the carcass had not plateaued even at 26% crude protein. These authors gave no indication of specific amino acid levels or the balance of amino acids.

In some of the experiments, where additional lysine was

added, the balance of amino acids would appear to be less than recommended for optimum performance. For example, in the experiments reported by Evock et al. (1988), the addition of lysine to the basal diet could have resulted in methionine plus cystine or threonine becoming the first limiting amino acid, on the basis that they were reduced to 50% of the lysine level. Boyd et al. (1988) suggested that protein intake may have limited a further response in one treatment where protein accretion reached 155 $g.pig^{-1}.d^{-1}$. They did not however state the protein level fed.

The growth response to pST is independent of feed level (Campbell et al., 1987; Campbell et al., 1988: Bechtel et al., 1988). However, in the latter study, there were indications that the carcass response was lower in pigs fed a restricted scale. This suggests that the carcass improvement response to pST may be greater on high energy diets. The diets which Campbell et al. (1988) used had a digestible energy content of 3.5 Mcal/kg and 1.2% lysine. Pigs were fed ad libitum, 80% of ad libitum and 60% of ad libitum over the liveweight phase 25 to 55 kg.

That rapid growth of muscle and bone tissue places greater demands on mineral and vitamin nutrition is suggested by the observation of mobility problems (Bryan et al., 1987; Evock et al., 1988). However, the latter recorded no differences in serum or bone calcium and phosphorus concentrations. Caperna et al. (1987) noted that bone ash content of pigs, treated with 100 $ug.kg\ BW^{-1}.d^{-1}$ ppST, was lower than that of control pigs. Haematocrit values were also lower in treated animals, as were iron concentrations in serum and liver tissue. The latter effects were, to some extent, influenced by feed intake in that they were less obvious with ad libitum fed pigs. Goff et al.

(1988) observed no differences in bone ash concentration, but whole body ash and carcass ash increased by 10 and 15% respectively, when pigs were injected with 100 ug ppST.kg $BW^{-1}.d^{-1}$. Furthermore there was an increased rate of vitamin D metabolism in treated pigs.

Environment for pST treated pigs

Reported experiments on pST treatment are generally based on small numbers of pigs, which for the most part appear to have been individually fed. There are no reports of treatment under what might be termed commercial conditions, primarily because of the treatment method available. However since pST reduces feed intake, one would expect little behavioral change on eating pattern and hence, existing feeding and housing systems would appear to be perfectly adequate for treated animals.

Because of the increased heat production of treated animals perhaps there should be some experimentation on climatic environment, especially temperature, in the case of pigs penned in large groups. For example van der Hel et al. (1988) recorded elevated heat production, 12 kcal.kg.$^{-.75}$ in treated pigs. This represented an increase of 8% over untreated pigs.

The improved feed utilization efficiency of treated pigs is estimated to reduce total effluent production by about 5%. Thus, pST would have a positive effect on other aspects of house environment and also on the ecological environment.

SUMMARY AND CONCLUSION

The literature quoted in this review originated primarily in North America. Many of the references are abstracts of papers presented at the American Society of Animal Science meetings and refer to recent research. It is obvious that a considerable amount of work is being undertaken on the topic and that within the near future the available literature will expand greatly.

pST, when administered to pigs between 50 kg liveweight and slaughter, results in major improvements in growth performance and carcass composition. The best response appears to be obtained with a dose of 30 to 60 ug.kgBW^{-1}.d^{-1}. This results in a c.5% reduction in feed intake, a c.20% improvement in growth rate and a c.20% reduction in the quantity of feed required per kg gain. The main changes in carcass composition are, a c.30% reduction in fat content and a c.10% increase in the lean content. All experiments do not report responses of this magnitude but those with a lesser response in general bear an explanation.

That no major alteration to diet is required to achieve the growth and carcass response is to some extent surprising, but understandable. Treatment with pST is simply an addition of a naturally occurring controlling hormone. However, because of the particularly large increase in muscle and bone tissue of pST-treated pigs, there is no conflict in expecting that some minor refinements are required in the balance of amino acids, minerals and vitamins presented. Feed or energy level may also influence the magnitude of response. These issues require further experimentation as indicated by variation in growth performance and carcass improvement and by the mobility problems reported in 2 studies.

With one exception all experiments reported administration of pST by daily intramuscular injection. This is not a practical method. With the reporting of a successful slow release ear implant however, this is no longer a matter of concern. Some experimentation on the environmental requirements of treated pigs is necessary, based on the observation that such pigs have a higher metabolic heat output.

REFERENCES

Bechtel, P.J., Easter, R.A., Novakofski, J., McKeith, F.K., McLaren, D.G., Jones, R.W. and Ingle, D.L. 1988. Effect of porcine somatotropin on limit fed swine. J. Anim. Sci. 66, (Suppl. 1): 282 (Abstr).

Boyd, R.D., Bauman, D.E., Beermann, D.H., De Neergaard, A.F., Souza, L. and Butler, W.R. 1986. Titration of the porcine growth hormone dose which maximises growth performance and lean deposition in swine. J. Anim. Sci. 63, (Suppl. 1): 218 (Abstr).

Boyd, R.D., Beerman, D.H., Roneker, K.R., Bartley, T.D. and Fagin, K.D. 1988. Biological activity of a recombinant variant (21 kd) of porcine somatotropin in growing swine. J. Anim. Sci. 66, (Suppl. 1): 256 (Abstr).

Bryan, K.A., Carbaugh, D.E., Clark, A.M., Hagen, D.R. and Hammond, J.M. 1987. Effect of porcine growth hormone (pGH) on growth and carcass composition of gilts. J. Anim. Sci. 65, (Suppl. 1): 244 (Abstr).

Bryan, K.A., Clark, A.M. and Hagen, D.R. 1988. Effect of treatment with and withdrawal of porcine growth hormone (pGH) on growth and reproductive performance of gilts. J. Anim. Sci. 66, (Suppl. 1): 401 (Abstr).

Campbell, R.G., Caperna, T.J., Steele, N.C. and Mitchell, A.D. 1987. Effect of porcine pituitary growth hormone (pGH) administration and energy intake on growth performance of pigs from 25 to 55 kg body weight. J. Anim. Sci. 65, (Suppl. 1): 244 (Abstr).

Campbell, R.G. and Travener, M.R. 1988. Genotype and sex effects on the responsiveness of growing pigs to exogenous porcine growth hormone (pGH) administration. J. Anim. Sci. 66, (Suppl. 1): 257 (Abstr).

Campbell, R.G., Steele, N.C., Caperna, T.J., McMurtry, J.P., Solomon, M.B. and Mitchell, A.D. 1988. Interrelationships between energy intake and endogenous porcine growth hormone administration on the performance, body composition and protein and energy metabolism of growing pigs weighing 25 to 55 kilograms liveweight. J. Anim. Sci. 66: 1643-1655.

Caperna, T.J., Campbell, R.G. and Steele, N.G. 1987. Influence of porcine growth hormone (pGH) administration and energy intake on mineral status of growing swine. J. Anim. Sci. 65 (Suppl. 1): 254 (Abstr).

Chung, C.S., Etherton, T.D. and Wiggins, J.P. 1985. Stimulation of swine growth by porcine growth hormone. J. Anim. Sci. 60: 118-130.

Cohen, S.N., Chang, A.C.Y., Boyer, H.W. and Helling, R.B. 1973. Construction of biologically functional bacterial plasmids in vitro. Proc. Nat. Acad. Sci. U.S.A. 70: 3240.

Daughaday, W.H. 1985. Prolactin and growth hormone in health and disease. In: Contemporary Endocrinology Vol. 2 Ed. S.H. Ingbar. Plenum Publ. Co., New York, N.Y. 27-86.

Etherton, T.D., Evock, C.M., Chung, C.S., Walton, P.E., Sillence, M.N., Magri, K.A. and Ivy, R.E. 1986. Stimulation of pig growth performance by long term treatment with pituitary porcine growth hormone (pGH) and a recombinant pGH. J. Anim. Sci. 63 (Suppl. 1): 219 (Abstr).

Etherton, T.D., Wiggins, J.P., Chung, C.S., Evock, C.M., Redhun, J.F. and Walton, P.E. 1986. Stimulation of pig growth performance by porcine growth hormone and growth hormone releasing factor. J. Anim. Sci. 63: 1389-1399.

Etherton, T.D., Wiggins, J.P., Evock, C.M., Chung, C.S., Rebhun, J.F., Walton, P.E. and Steele, N.C. 1987. Stimulation of pig growth performance by porcine growth hormone: Determination of the dose-response relationship. J. Anim. Sci. 64: 433-443.

Evans, F.D., Osborne, V.R., Evans, N.M., Morris, J.J. and Hacker, R.R. 1988. Effect of different patterns of administration of recombinant porcine somatotropin (rpST) to pigs from 5 to 15 weeks of age. J. Anim. Sci. 66, (Suppl. 1): 256 (Abstr).

Evans, H.M. and Simpson, M.E. 1931. Hormones of the anterior hypophysis. Amer. J. Physiol. 98: 511-546.

Evock, C.M., Etherton, T.D., Chung, C.S. and Ivy, R.E. 1988. Pituitary porcine growth hormone (pGH) and a recombinant pGH analogue stimulate pig growth performance in a similar manner. J. Anim. Sci. 66: 1928-1941.

Giles, D.D. 1942. An experiment to determine the effect of growth hormone of the anterior lobe of the pituitary gland on swine. Amer. J. Vet. Res. 3: 77-86.

Goff, J.P., Caperna, T.J., Campbell, R.G. and Steele, N.C. 1988. Interactions of porcine growth hormone (pGH) administration and dietary energy intake on circulating vitamin D metabolite concentrations in growing pigs. J. Anim. Sci. 66, (Suppl. 1): 291 (Abstr).

Henricson, B. and Ullberg, S. 1960. Effect of pig growth hormone on pigs. J. Anim. Sci. 19: 1002-1008.

Huisman, J., van Weerden, E.J., van der Hal, W., Verstegen, M.W.A., Kanis, E. and van den Wal, P. 1988. Effect of rpST treatment on rate of gain in protein and fat in two breeds of pigs and as crossbreds. J. Anim. Sci. 66, (Suppl. 1): 254 (Abstr).

Ivy, R.E., Baldwin, C.D., Wolfrom, G.W. and Mouzin, D.E. 1986. Effect of various levels of recombinant porcine growth hormone (rpGH) injected intramuscularly in barrows. J. Anim. Sci. 63, (Suppl. 1): 218 (Abstr).

Kanis, E., van der Hel, W., Huisman, J., Nieuwhof, G.J., Politiek, R.D., Verstegen, M.W.A., van der Wal, P. and van Weerden, E.J. 1988. Effect of recombinant porcine somatotropin (rpST) treatment on carcass characteristics and organ weights of growing pigs. J. Anim. Sci. 66, (Suppl. 1): 280 (Abstr).

Knight, C.D., Azain, M.J., Kasser, T.R., Sabacky, M.J., Baile, C.A., Buonomo, F.C. and McLaughlin, C.L. 1988. Functionality of an implantable 6 week delivery system for porcine somatotropin (pST) in finishing hogs. J. Anim. Sci. 66, (Suppl. 1): 257 (Abstr).

Kraft, L.A., Haines, D.R. and DeLay, R.L. 1986. The effect of daily injections of recombinant porcine growth hormone (rpGH) on growth, feed efficiency, carcass composition and selected metabolic and hormonal parameters in finishing swine. J. Anim. Sci. 63, (Suppl. 1): 218 (Abstr).

Laron, Z. 1983. Deficiencies of growth hormone and somatomedins in man. Special topics in endocrinology and metabolism, 5: 149-199.

Machlin, L.J. 1972. Effect of porcine growth hormone on growth and carcass composition of the pig. J. Anim. Sci. 35: 794-800.

McLaren, D.G., Grebner, G.L., Bechtel, P.J., McKeith, F.K., Novakofski, J.E., Easter, R.A., Jones, R.W. and Dalrymple, R.H. 1987. Effect of graded levels of natural porcine somatotropin (PST) on growth performance of 57 to 103 kg pigs. J. Anim. Sci. 65, (Suppl. 1): 245 (Abstr).

McLaughlin, C.L., Baile, C.A., Qui, S-Z and Wang, L-C. 1988.

Responses of Beijing Black hogs to porcine somatotropin (PST) treatment. J. Anim. Sci. 66, (Suppl 1): 255 (Abstr).

Newcomb, M.D., Grebner, G.L., Bechtel, P.J., McKeith, F.K., Novakofski, J., McLaren, D.G., Easter, R.A. and Jones, R.W. 1988. Response of 60 to 100 kg pigs treated with porcine somatotropin to different levels of dietary crude protein. J. Anim. Sci. 66, (Suppl. 1): 281 (Abstr).

Rebhun, J.F., Etherton, T.D., Wiggins, J.P., Chung, C.S., Walton, P.E., and Steek, N. 1985. Stimulation of swine growth performance by porcine growth hormone (pGH): Determination of the maximally effective pGH dose. J. Anim. Sci. 61, (Suppl. 1): 251 (Abstr).

Turman, E.J. and Andrews, F.M. 1955. Some effects of purified anterior pituitary growth hormone on swine. J. Anim. Sci. 14:7-19.

van der Hel, W., Verstegen, M.W.A., Huisman, J., Kanis, E., van Weerden, E.J. and van der Wal, P. 1988. Effect of rpST treatment on energy balance traits and metabolic rate in pigs. J. Anim. Sci. 66, (Suppl. 1): 225 (Abstr).

Wolfrom, G.W., Ivy, R.E. and Baldwin, C.D. 1986. Effect of native porcine growth hormone (npGH) injected intramuscularly in barrows. J. Anim. Sci. 63, (Suppl. 1): 219 (Abstr).

MILK FROM BST-TREATED COWS; ITS QUALITY AND SUITABILITY FOR PROCESSING

G. van den Berg

NIZO (Netherlands Institute for Dairy Research)
P.O. Box 20, 6710 BA Ede, The Netherlands

ABSTRACT

This paper deals with the question if cows injected with bovine somatotropin still give milk of good quality and processing characteristics.
The composition of milk is very complex. Therefore, it is necessary that not only individual milk components are analysed, but also various functional properties are studied. This matter is discussed before reviewing the recent information on the subject. Fat, protein and lactose contents of the milk are not affected in a status of positive energy balance. With the sustained release system temporarily this balance may be negative and some cyclic effects in fat and protein contents are found. The protein composition is not affected, so neither the theoretical cheese yield calculated from the total protein content. The available information on various minerals and enzymes in milk gives no ground for uneasiness and neither do freezing point, pH-value and milk flavour. Too little is known about the effect on the various vitamins, and there is still some concern on the somatic cell count. Heat stability of the (condensed) milk and cheesemaking properties are not significantly affected, but more experiments are recommended. The level of growth hormone in the milk is not found to be affected by normal doses. The presence of an insulin-like growth factor in the milk asks some more attention.

The complexity of milk

Milk is an important food for direct consumption and is used also for the manufacture of a wide range of products. It has a very complex composition with unique biophysical properties, which can easily be disturbed. Nevertheless, the cow can be pushed by injecting bovine somatotropin to produce more milk with the same mammary gland and it seems still to remain the normal product.
However, milk is also subject to many processing technologies, and to check its properties with respect to that, it is not sufficient only to determine the contents

of various milk constituents. The combination of many
components and their concentrations determine if the
suitability of the milk for processing has been changed.
Therefore various processing characteristics are to be
determined as such in addition to certain chemical
parameters.

With respect to the different processes in which milk is
utilized roughly three categories of milk properties can
be distinguished as mentioned in Table 1.

TABLE 1   Categories of milk properties.

Quality	sensoric characteristics microbiological contaminations chemical contaminations
Composition	processing characteristics product yield nutritive value
Physical/biological properties	processing characteristics fermentation properties product yield

Milk ought to possess the desired organoleptical quality
with a minimal microbial contaminating flora, without
chemical contaminants and pathogenic micro-organisms. The
nutritive value is directly dependent on composition, as is
product yield. With respect to milk composition one example
is mentioned: calcium content, pH-value, etc. have a strong
influence on the stability of milk and its renneting
properties. The colloidal properties of the milk determine
many processing characteristics and should not deviate from
their normal state. Fermentation properties depend on many
minor components. When such properties are not normal
process control during manufacturing is more difficult and a
lower yield or a low product quality may be the result.

Gross composition

The influence of the BST-treatment on the fat content,
the total protein content and the lactose content of the

milk has been investigated very extensively. That is well reviewed by McBride et al (1988). A daily administration of exogenous growth hormone derived from the extracts of pituitary glands of slaughtered cows or recombinantly derived somatotropin causes a higher milk yield, while the gross composition does not change. A prerequisite is a ration of full value that has to be taken by the cow in sufficient quantities to avoid a negative status of the energy balance. Otherwise the fat content of the milk will increase and the protein content may decrease. This may be concluded from short- and long-term experiments. In this respect a good farm management is necessary, which can be stimulated by a modern pricing system for the farm milk by the dairy industry.

It may be expected that a biweekly or four weekly administration in a sustained release vehicle is more practicable. Effects of this technique have been reviewed by Rohr (1988). On average the milk has fat and protein contents similar to those of milk from untreated cows. However, a cyclic effect on yield and composition was observed. The first week after treatment of the cows the fat content rose together with the milk yields, and both decreased afterwards (Oldenbroek et al, 1987; Farries and Profittlich, 1988). The protein content increased slightly during the fourth week after injection. Such effects may be smaller with a higher treatment frequency. A short time of a negative energy balance seems to cause these effects. This way of administration, if not improved on this point and when utilized on a farm where cheese is manufactured, will make the cheese process control more complicated. Treatment on a number of farms within a few days may have some influence on the milk processing in the plant in spite of the milk being standardized on fat content.

The increase in fat content is caused by a mobilization of the fat reserves of the animal and lead to an increase of the longer unsaturated fatty acids in the milk fat. This makes the fat softer which was notable by a slightly improved spreadability of the butter (Gravert, 1988).

Protein composition

The proteins form a very important group of milk constituents because of their impact on milk properties and product yield and their high nutritive value. The casein content is important for cheese yield and casein (product) yield. Besides, the milk price is often based partly on the total protein content. This makes it important not to decrease the casein-to-total protein ratio. A compilation of data on this topic is given in Table 2.

TABLE 2 Effect of BST-treatment on content and composition of milk proteins. (↓decreased; ↗increased; = similar or n.s.)

| Reference | Casein | | Whey proteins | | Casein/ |
	Content	Composition	Content	Composition	Total protein
Leonard et al 1988	=	=	=		=
Baer et al 1988	=		=		=
Auberger et al 1988[1]	=	=	=		=
Desnouveaux et al 1988[1]	=		=		=
Kindstedt et al 1988	=		↗		↓
Lynch et al 1988b		=		3)	
Vignon 1987	=	=	=	=	=
Vignon 1988[2]	=	=	=	=	=
Vignon and Ramet 1988[2]	=	=	=		=
Van den Berg and De Jong 1986	=		=		=
Escher and Van den Berg 1987	↓		↓		=4)
Escher and Van den Berg 1988b	=		=		=
Pabst et al 1987b	=	=	=	=	=

1) milk from the same herd
2) milk from the same herd
3) α-lactalbumine increased
4) fat content increased/protein content decreased; probably a negative energy balance at the time of this test.

Generally, the casein content is not markedly influenced as little as the casein composition. That is to say that the ratio between the different caseins in the micelles is not affected nor are the calcium, phosphate and citrate in the micelle. This is very important for the casein properties during processing. The whey protein content is not much affected either, nor is the casein-to-total protein ratio. Unfortunately, there are only a few observations on the individual whey proteins.

Minerals

Since the cation flux from blood plasma into the mammary gland and into the milk is increased with the milk yield in order to maintain osmoregulation, the concentration of the minerals is not affected (McBride et al, 1988). This is confirmed for Ca, P, Na, Fe, Cu and Mn (Eppard et al, 1985). These findings for Ca and P came also from other investigations even in cases with high overdoses of BST (Hard et al, 1988 and Wilkinson, 1988). Also for Zn no effect was found (Wilkinson, 1988). For total Ca, soluble Ca, ionic Ca, micellar Ca, P, Na, K and Mg a similar result was observed (Desnouveaux et al, 1988). No difference in total Ca was found either by Auberger et al (1988).

Enzymes

Little research has been conducted into the enzymes in milk in relation to the administration of BST to the dairy-cow. An increased proteolytic activity in the milk may prematurely break down the casein and might decrease the cheese yield. Such enzymes can also pass into the cheese and affect the ripening process. An investigation on the aminopeptidase and endopeptidase activities in the milk showed no significant differences between experimental and control milk (Desnouveaux et al, 1988). In the same experiments, no difference in phosphatase activity was found. Protease activity was checked by Lynch et al (1988a), but no difference caused by the treatment was found either. This study does not mention the type of protease investigated, but because of the increase of the activity with the stage of lactation, it probably concerned plasmin. The lipase activity was also checked and no significant differences were found in the same experiments. Moreover, when the dairy product properties are treated later on it will also be clear that negative effects by increased enzyme activities are not likely.

Vitamins

On this subject the only information is coming from a Monsanto Report (Hartnell, 1986). Vitamin A, thiamine, riboflavin, pyridoxine, vitamin B12, biotin, panthotenic acid and cholin have been investigated. There were no differences except for biotin that was present in a slightly increased concentration in the BST-milk. More research on this subject is recommended.

Pyruvate

In the Federal Republic of Germany milk quality is often checked by means of its pyruvate content. A higher value was found in the milk from BST-treated cows (Gravert, 1988). Should this appear a general tendency in BST-milk, then it might make this checking method difficult to apply.

Freezing point

The freezing point of the milk is not significantly influenced, which can be concluded from a few investigations (Desnouveaux et al, 1988, Van den Berg and De Jong, 1986, Escher and Van den Berg, 1987 and Escher and Van den Berg, 1988b). There is a certain relationship between freezing point and solids-not-fat content; in the last mentioned reference such a tendency has been demonstrated.

pH-Value

The pH-value of the milk was not influenced as has been found in five investigations (Van den Berg and De Jong, 1986, Escher and Van den Berg, 1987, Battistotti and Bertoni, 1988 Vignon, 1988 and Vignon and Ramet, 1988).

Milk flavour

In three investigations no significant difference in flavour was found between the fresh milk from treated and untreated cows (Baer et al, 1988, Van den Berg and De Jong, 1986 and Escher and Van den Berg, 1987). After ten days of cold storage the milk was still of good quality (Escher and Van den Berg, 1987) and it was also reported that the milk was not susceptible to oxidation flavour (Baer et al, 1988). Such an off-flavour could come forward during cold storage.

Somatic cell count

This point is part of the milk quality control system in the whole European community. From eight experiments no significant differences caused by the BST-treatment are reported (Enright et al, 1988, Rowe-Bechtel et al, 1988, Munneke et al, 1988, Aguilar et al, 1988, Hart et al, 1988, Oldenbroek et al, 1987, Kindstedt et al, 1988 and Wilkinson, 1988). A slight increase on an acceptable level was found in two cases (Gravert, 1988 and Pell et al, 1988). At doses of BST much more than ever will be practised a high count was found in one case (McBride et al, 1988), and in another trial an increased incidence of mastitis was reported (Eppard et al, 1988). Although practical doses of BST may give no reason for major concern, this topic should be checked in further experiments under European conditions.

Stability of the milk

The alcohol stability of the milk was normal in three tests (Van den Berg and De Jong, 1986, Escher and Van den Berg, 1987 and Escher and Van den Berg, 1988b). This functional property of the milk was also checked by heat coagulation experiments. No significant difference in coagulation time of milk was found (Auberger et al, 1988). But heat stability is more critical after concentration when producing condensed milk. This has been checked also with a good result (Van den Berg and De Jong, 1986 and Escher and Van den Berg, 1987). Heat stability is supposed to have some relationship with the urea content of the milk. Urea is an important part of the non-protein-nitrogen (NPN). In this respect the NPN-content of the milk might be of interest.

From the information available about the nitrogen compounds in milk no large differences in NPN may be expected, but from two experiments an increase in NPN has been reported (Bauman et al, 1988 and Kindstedt et al, 1988). However, it would be better to estimate the urea content itself with respect to heat stability.

Cheesemaking properties

The renneting reaction (as a basic part of the cheese manufacturing process) is very sensitive to check deviations in some milk properties. In addition, the genetic code of the cow is imperative to the properties of the caseins that take part in this reaction. Therefore when larger herds are used for the experiments, differences by polymorphism of the proteins between the groups of cows are less probable. The results in reports on cheesemaking from milk from treated cows versus untreated ones are summarized in Table 3.

The last two experiments were probably influenced by the rather small group of cows giving room for differences in milk properties caused by indivudual cows and included milk with different protein contents. This resulted in different syneretic behaviour and moisture content which influence also the sensoric properties of the cheese. But the general impression gained from this Table is that the cheesemaking properties are generally not affected by the BST-treatment of the cows. However, several cheesemaking experiments are carried out with milk from a more or less arbitrary moment of time during a long-term farm experiment.

An experiment is still running in the United Kingdom where Cheddar and Wensleydale cheese were made. It was reported that the cheesemaking process and the composition of the cheeses were similar for milk from control and experimental groups (Phipps, 1988). Repeated tests during several lactation stages of bigger herds are desirable.

TABLE 3 Effect of BST-treatment on cheese manufacture
(↓decreased; ↗ increased; = similar or n.s. # different)

Reference	Coagu-lation	Acidifi-cation	Syn-eresis	Cheese composition	Sensoric properties	Cheese type
Pabst et al 1987a	=	=	=	=	=	?
Auberger et al 1988[1]	=		=			
Desnouveaux et al 1988[1]	=	=	=	=	=	Camembert
Vignon 1988[2]	=	=	=			
Vignon and Ramet 1988[2]	=	=	=	=	=	Camembert St. Paulin
Vassal 1988[2]	=		=		=[3]	Camembert St. Paulin
Battistotti and Bertoni 1988	=/↓[4]	=				none
Van den Berg and De Jong 1986[6]	=	=	=	=	=	Gouda
Escher and Van den Berg 1987[7]	↓	=	↗	#	#	Gouda
Escher and Van den Berg 1988b[7]	=	=	↓	#	=[5]	Gouda

1) Milk from the same herd
2) Milk from the same herd
3) First experiment St. Paulin different
4) At end of lactation some treated cows gave milk with somewhat poorer renneting properties
5) slight differences in moisture content; after 6 weeks of maturation no difference; after 3 months of maturation one cheese deviated because of slight butyric acid fermentation
6) milk from the herd under investigation by Rypkema et al 1987
7) milk from the herd under investigation by Oldenbroek et al 1987 (9 and 14 treated animals respectively)

Excretion of hormones in milk

The concentration of natural bovine somatotropin in milk is approximately at the detection level of 0.3-0.5 ng/ml. Some studies are dedicated to the effect of the administration of exogenous BST to the cow on this content. Normal doses have no significant effect on the level in milk (Torkelson et al, 1987, Hart et al, 1985, Heeschen, 1988 and Schams, 1987). Only very high doses give a measurable increase (Hart et al, 1985 and Schams, 1987).

Other hormones involved in the physiological system of the cow and influenced by the growth hormone are the insulin-like growth factors I and II (somatomedins). Their presence in the milk is increased by the BST-administration to the cow, albeit still at concentration levels within the range of natural values (Torkelson et al, 1988 and Collier et al, 1988). There is a strong similarity with the human insulin-like growth factors (Honegger and Humbel, 1986 and Francis et al, 1986) which give ground for questioning its effect on the consumer. More investigations on the amount of these growth factors in milk and on the evidence of their harmlessnes are desirable.

Nutritive value

Because of the very small deviations, if any, the nutritive value of the milk from treated cows need not be discussed. Only a little more information on vitamins could be valuable, but the main compounds that make milk and dairy products to a very valuable food are not at stake.

Concluding remarks

High-producing cows have a high level of growth hormone in their blood plasma. Administration of bovine somatotropin (or growth hormone releasing factor) makes any cow similar to a high-producing one. In this respect, it seems to be obvious that the treatment of the cow with recombinantly derived bovine somatotropin does not affect the wholesomeness and the processability of the milk. On the farms with such dairy cows a good management is required. At least it is necessary to supply the cow with a good ration to protect them against negative energy balances.

But, when such far-reaching interventions as the introduction of BST are possible many aspects should be considered and seriously investigated. In several countries such investigations on milk quality are still continuing.

Some experiments were carried out with rather small herds so that other factors may have influenced the results. It is advisable to repeat the milk processing experiments a number of times with milk from herds that are selected also genetically.

Some more information on the effect on vitamins is desirable.

The level of the insulin-like growth factors (IGF-I and II) should be checked more thoroughly and severely considered. Also the number of the somatic cell count should be paid some more attention.

If the BST-treatment is authorized, the system for prolonged release should be improved to avoid cyclic effects as far as possible.

REFERENCES

Aguilar, A.A., Jordan, D.C., Olson, J.D., Baily, C. and Hartnell, G. 1988. A short-term study evaluating the galactopoietic effects of the administration of Sometribove (recombinant methionyl bovine somatotropin) in high producing dairy cows milked three times per day. J. Dairy Sci., 71(suppl. 1),208.

Auberger, B., Lenoir, J. and Remeuf, F. 1988. L'incidence du traitement de vaches laitières par la somatotropine bovine sur la composition et les aptitudes technologiques du lait. Techn. Lait. Mark., 1030, Technologie 1-3.

Baer, R.J., Treszen, K.M., Schingoethe, D.J., Caspar, D.P., Shaver, R.D. and Cleale, R.M. 1988. Composition and flavour of milk produced by cows injected with recombinant bovine somatotropin. J. Dairy Sci., 71 (suppl. 1)115.

Battistotti, B. and Bertoni, G. 1988. Preliminary report on the results of analyses of milk from cows treated with BST and untreated cows. Catholic University Sacred Heart, Fac. Agric., Piacenza.

Bauman, D.E., Hard, D.L., Crooker, B.A., Erb, H.N. and Sandles, L.D. 1988. Lactational performance of dairy cows treated with a prolonged-release formulation of methionyl bovine somatotropin (sometribove). J. Dairy Sci., 71 (suppl. 1)205.

Collier, R.J., Li, R., Johnson, H.D., Becker, B.A., Buonomo, F.C. and Spencer, K.J. 1988. Effect of sometribove (metionyl bovine somatotropin, BST) plasma insulin-like growth factor I, (IGF-I) and II, (IGF-II) in cattle exposed to heat and stress. J. Dairy Sci., 71(suppl. 1) 228.

Desnouveaux, R., Montigny, H., Le Treut, J.-H., Schockmel, L. and Biju-Duval, B. 1988. Vérification de l'aptitude fromagère du lait de vaches traitées à la somatotropine bovine méthionylée: Fabrication expérimentale de fromages a pâtes molles de type camembert. Techn. Lait. Mark., 1030, Technologie 4-7.

Enright, W.J., Chapin, L.T., Mosely, W.M. and Tucker, H.A. 1988. Effects of infusions of various doses of bovine growth hormone-releasing factor on growth hormone and lactation in Holstein cows. J. Dairy Sci., 71,99-108.

Eppard, P.J., Bauman, D.E., Bitman, J., Wood, D.L., Akers, R.M. andd House, W.A. 1985. Effect of dose of bovine growth hormone on milk composition: α-lactalbumin, fatty acids and mineral elements. J. Dairy Sci., 68, 3047-3054.

Eppard, P.J., Lanza, G.M., Hudson, S., Cole, W.J., Hintz, R.L., White, T.C., Ribelin, W.E., Hammond, B.G., Bussen, S.C., Leak, R.K. and Meztger, L.E. 1988. Response of lactating dairy cows to multiple injections of sometribove, USAN(recombinantly methionyl somatotropin) in a prolonged release system. Part I. Production response. J. Dairy Sci., 71(suppl. 1)184.

Escher, J.T.M. and Van den Berg, G. 1987. Investigations on the influence of r-DNA bovine somatotropine on milk properties. Report NOV-1245, Neth. Inst. Dairy Res., Ede.

Escher, J.T.M. and Van den Berg, G. 1988a. The effect of somidobove in a slow release vehicle on some milk constituents and the renneting properties. Report NOV-1309, Neth. Inst. Dairy Res., Ede.

Escher, J.T.M. and Van den Berg, G. 1988b. The effect of somidobove in a sustained release vehicle on some milk constituents and cheesemaking properties. Report NOV-1316, Neth. Inst. Dairy Res., Ede.

Farries, E. and Profittlich, C. 1988. Veränderung einiger stofwechselparameter bei der Milchkuh. BST-Symposium. Landbauforschung Völkenrode, Sonderheft 88,135-158.

Francis, L. Read, C., Ballard, F.J., Bagley, C.J., Upton, F.M., Gravestock, M. and Wallace, J.C., 1986. Purification and partial sequence analysis of insulin-like growth factor-1 from bovine colostrum. Biochem. J.,223,207-213.

Gravert, H.O. 1988. Behandlung von Kühen mit gentechnologisch gewonnenem Wachstumshormon. DMZ Welt der Milch,42, 917-918.

Hard, D.L., Cole, W.J., Fransen, S.E., Samuels, W.A., Bauman, D.E., Erb, H.N., Huber, J.T. and Lamb, R.C. 1988. Effect of long term sometribove, USAN(recombinant methionyl bovine somatotropin), treatment in a prolonged release system on milk yield, animal health and reproductive performance-pooled across four sites. J. Dairy Sci.,71, (suppl. 1)210.

Hart, I.C., Bines, J.A. and Morant, S.V. 1985. The effect of injecting or infusing low doses of bovine growth hormone on milk yield, milk compositon and the quantity of hormone in the milk serum of cows. Anim. Prod., 40, 243-250.

Hartnell, G. E. 1986. Evaluation of vitamins in milk produced from cows treated with placebo and CP115099 in a prolonged release system. Monsanto Technical Report MSL-7031.

Heeschen, W. 1988. bST in der Milch. BST-Symposium. Landbauforschung Völkenrode, Sonderheft 88, 220-226.

Honegger, A. and Humbel, R.E. 1986. Insulin-like growth factors I and II in fetal and adult bovine serum. J. Biol. Chem., 261, 569-575.

Kindstedt, P.S., Rippe J.K., Pell, A.N. and Hartnell, G.F. 1988. Effect of long-term administration of sometribove, USAN (recombinant methionyl bovine somatotropin) in a prolonged release formulation on protein distribution in Jersey milk. J. Dairy Sci., 71,(suppl. 1)96.

Leonard, M., Turner, J. and Block, E. 1988. Effects of long-term somatotropin injection in dairy cows on milk protein profiles. Res. Rep., Dept. Animal Sci., Mc Gill University, Quebec, Canada. Ed. H. Garino, january 1988.

Lynch, J.M., Senyk, G.F., Barbane, D.M., Bauman, D.E. and Hartnell, G.F. 1988a. Influence of sometribove (recombinant methionyl bovine somatotropin) on milk lipase and protease activity. J. Dairy Sci., 71(suppl. 1) 100.

Lynch, J.M., Barbano, D.M., Bauman, D.E. and Hartnell, G.F. 1988b. Influence of sometribove (recombinant methionyl bovine somatotropin) on the protein and fatty acid composition of milk. J. Dairy Sci., 71(suppl. 1) 100.

McBride, B.W., Burton, J.L. and Burton, J.H. 1988. The influence of bovine growth hormone (somatotropin) on animals and their products. Res. Developm. Agric., 5(1), 1-21.

Munneke, R.L., Sommerfeldt, J.L. and Ludens, E.A. 1988. Lactational responses of dairy cows to recombinant bovine somatotropin. J. Dairy Sci., 71(suppl. 1) 206.

Oldenbroek, J.K., Garssen, G.J., Forbes, A.B. and Jonker, L.J. 1987. The effect of treatment of dairy cows of different breeds with recombinantly drived bovine somatotropin in a sustained delivery vehicle. 38th Ann. Meeting EAAP, Lisbon.

Pabst, K., Prokopek, D., Peters, K.H. and Krusch, U. 1987a. Effect of application of BST on technological properties of milk and quality of products. Fed. Dairy Res. Centre, Kiel, Fed. Rep. Germany, Ann. Rep. 1987, B8.

Pabst, K., Roos, N. and Sick, H. 1987b. Influence of BST-treatment of cows on milk protein pattern. Fed. Dairy Res. centre, Kiel, Fed. Rep. Germany, Ann. Rep. 1987, B9.

Pell, A.N., Tsang, D.S., Huyler, M.T., Howlett, B.A. and Kunkel, J. 1988. Responses of Jersey cows to treatment with sometribove, ASAN(recombinant methionyl bovine somatotropin) in a prolonged release system. J. Dairy Sci.,71, (suppl. 1) 206.

Phipps, R.H. 1988. Pers. comm.

Rohr, K. 1988. Milchleistung und Milchzusammensetzung. BST-Symposium. Landbauforschung Völkenrode, Sonderheft 88, 104-115.

Rowe-Bechtel, C.L., Muller, L.D., Deaver, D.R. and Griel, L.C. Jr. 1988. Administration of recombinant bovine somatotropin (rbSt) to lactating dairy cows beginning at 35 and 70 days postpartum. I. Production response. J. Dairy Sci., 71(suppl. 1) 166.

Rypkema, Y.S., Van Reeuwijk, L. Peel, C.J. and Mol, E.P. 1987. Responses of dairy cows to long-term treatment with somatotropin in a prolonged release formulation. 38th Ann. Meeting EAAP, Lisbon.

Schams, D. 1987. Analytik Wachstumhormon. Analytik des endogenen und exogenen Wachstumshormons beim Rind. Süddeutsche Vers. Forsch. Anst. Milchwirtsch. Weihenstephan, Techn. Universität, München. Wissensch. Jahresbericht 1987.

Torkelson, A.R., Dwyer, K.A., Rogan, G.J. and Ryan, R.L. 1987. Radioimmunoassay of somatotropin in milk from cows administered recombinant bovine somatotropin. J. Dairy Sci., 70(suppl. 1) 146.

Torkelson, A.R., Lanza, G.M., Birmingham, B.K., Vicini, J.L., White, T.C., Dwyer, S.E., Madsen, K.S. and Collier, R.J. 1988. Concentration of insulin-like growth factor I (IGF-I) in bovine milk: Effect of herd, stage of lactation and sometribove, USAN (recombinant methionyl bovine somatotropin). J. Dairy Sci., 71(suppl. 1) 169.

Van den Berg, G. and De Jong, E. 1986. The influence of the treatment of lactating cows with Methionyl Bovine Somatotropin on milk properties. Report NOV-1209, Neth. Inst. Dairy Res., Ede.

Vassal, L. 1988. Fabrication de fromages de type Camambert et St. Paulin à partir de lait de vaches traitées(E) ou non (T) à la somatotropine, resumé des principales observations. Rapport préliminair, INRA, Stat. Rech. Lait., Jouy-en-Josas.

Vignon, C.S. 1987. "Injection of somatotropin to grazing dairy cows in 1986": Effect on the composition of fats and nitrogenous substances in milk. Concluding experimental report, Inst. Nat. Polytech. Lorraine, Nancy.

Vignon, C.S. 1988. "Effect of the injection of somatotropin in dairy cows on the composition and technological value of milk". End of experiment report, Inst. Nat. Polytech. Lorraine, Nancy.

Vignon, C.S. and Ramet, J.P. 1988. Effect on the technological value of milk by injecting dairy cows with somatotropin. End of experiment report, Inst. Nat. Polytech. Lorraine, Nancy.

Wilkinson, J.I.D. 1988. Pers. comm. on U.S. Field trials and one year target animal safety study. Eli Lilly Res. Centre, Ltd. Erlwood, UK.

SOMATOTROPIN AND RELATED PEPTIDES IN MILK

D. Schams

Institut für Physiologie
Südd. Versuchs- und Forschungsanstalt für Milchwirtschaft
Technische Universität München, 8050 Freising-Weihenstephan
FRG

ABSTRACT

Bovine somatotropin (bST) and insulin like growth factor-I (IGF-I) were measured in cow milk before and during treatment with a recommended dose (640 mg s.c.) of a sustained-release formulation of recombinant bST. In agreement with the literature no increase in milk somatotropin was detected compared to untreated cows; all concentrations were below 0.5 ng/ml. Only during the first days after the injection of a provocative dose (1.92 g s.c.) a small measurable increase was observed (maximal individual value 4.2 ng/ml). Concentrations of IGF-I in milk are much lower than in blood. A clear measurable increase was seen only after application of the provocative dose. The data presented and from the literature indicate that bST and IGF-I occur naturally in milk at low concentrations and that no clear increase was observed in cows, when administered bST in a sustained-release formulation according to dose recommendation.

INTRODUCTION

After supplementation of dairy cows with bovine somatotropin (bST) it may be supposed that bST and related peptides are secreted into milk and may be a potential risk for the consumer. Milk is composed of constituents or precursors from blood; accordingly hormones that are transported in blood may be detectable in milk. For the transport into milk passive and active mechanismen are discussed. Under physiological conditions in general steroid hormones (gestagens, oestrogens, corticoids and androgens) and protein hormones can be detected by bioassay and radioimmunoassay in milk in a variety of species (for review see Schams & Karg, 1986; Koldovsky & Thornburg, 1987). Determination of progesterone and oestrone sulfate in milk serves as a diagnostic tool in fertility control especially in cows. Data for protein hormones in milk are limited. Prolactin is of specific interest since its structure and moleculare size are similar to somatotropin. Prolactin in milk was measured in women, cattle, sheep, goat, sow and rat. Concentrations of prolactin in milk tend to be similar to those in blood plasma especially for long term secretions. Prolactin can be better measured in cow milk than in blood plasma due to stressfree (Schams & Schmidt-Polex, 1978) sample collection. The aim of the paper was to measure bST and IGF-I in milk before and during treatment with bST.

MATERIALS AND METHODS

Experiment 1; 8 Fleckvieh cows were used for this experiment (4 controls and 4 treated with a slow release bST preparation). Blood and milk were collected after the 3rd and 4th injection of bST (640 mg, s.c.) given in 4 week intervals. Blood was sampled by needle puncture from the jugular vein in tubes containing EDTA, chilled in ice water for 5 min and then centrifuged at 4°C. Plasma was stored at -20° C until analysed. Milk samples were cooled after collection for transport to the lab, then warmed up for 20 min at 38° C in a water bath, centrifuged upside-down for 30 min at 4500 RPM and skim milk was afterwards decanted and stored at -20°C until analysed. Blood and milk was sampled in the morning 1 day before the injection (-1) and further on day 1,3,6,8,1o,13,15,17,20,22,24,27 and after the fourth injection again on day 1,3,6,8,10 and 13.

Experiment 2; This experiment was done using 6 Brown Swiss cows. Blood and milk were collected as in exp. 1 on day 4, 1 and 0 before a single injection of a provocative dose of a slow release preparation of bST (1.92 g, s.c.) and afterwards on day 1,3,6,8,10,13, 15, 17, 20, 22, 24, 27, 29 and 31. All animals were fed with corn silage, hay and concentrate according to milk yield.

Hormone analysis

Somatotropin was determined in blood plasma by radioimmunoassay. The antiserum was raised in a rabbit using pituitary somatotropin (USDA-bGH-B-1) for immunization and was highly specific. No cross reactions with other anterior pituitary hormones were observed. A pure pituitary preparation (USDA-bGH-I-1) was used for iodination by the iodogen method. After separation of unlabelled from labelled STH by an anion exchange resin (A 62-X8; Bio Rad Lab., USA) further purification was achieved by column chromatography (Sephadex G-75). Standard or sample and antiserum were preincubated for 24 h at $4°$ C, then labelled bST was added and further incubated for 2 days. Separation of bound and free hormone was done with protein A (IgGsorb, Enzyme Center, Inc., Malden, MA, USA). Bovine pituitary STH (USDA-bGH-B-1, biological activity 1.9 IU/mg) served as reference preparation. The assay sensitivity was 0.5 ng/ml blood plasma or skim milk. The intra-assay CV estimated from two control samples running at the beginning, middle and end of each assay averaged 6.7 and 7.1 % respectively. The inter-assay coefficient of variation estimated from control samples of low, medium and high bST concentrations ranged from 8.4 - 12.5 %.

BST in milk was validated by parallelism of diluted milk samples with the standard curve and recovery experiments (adding cold bST to whole milk). Recovery was on average in 2 experiments 89 $\pm$ 12 % and 93 $\pm$ 17 % respectively.

Insulin-like growth factor (IGF-I) was analysed by RIA. Total IGF-I was measured after acid-ethanol extraction of 50 µl blood plasma or skim milk. Recombinant IGF-I (Amersham, England) was used for iodination according to the lactoperoxidase method. Synthetic IGF-I (kindly supplied by the late C.H. Li, San Francisco, USA) was used as reference preparation for calibration of a bovine plasma pool standard. The intra-assay variation and the inter-assay variation were on average 10 % and 14 % respectively.

RESULTS

The data (mean) for bST and IGF-I in blood and milk of experiment 1 are shown in Fig. 1. After both injections blood concentrations of bST increased clearly reached a maximum on day 3 and declined thereafter to basal levels on days 10 - 12. Values for the control group did not change significantly. Contrary to blood all samples in milk (controls and treatment) were below 0.5 ng/ml. IGF-I in blood increased from 264 ng/ml after the injection to a maximum on day 5 (1237 ng/ml) or 8 (784 ng/ml) and declined thereafter to basal levels around day 17. Concentrations of the control group fluctuated between 300 and 500 ng/ml. IGF-I levels in milk of controls are much lower than in blood and did not change significantly (range 26-33 ng/ml). Contrary to the blood concentrations treatment had no effect on IGF-I milk values. Levels tended to be a little higher (range 32-44 ng/ml) than in controls but the difference was not significant.

The results of experiment 2 injecting a provocative high dose of bST are shown in Figure 2. Concentrations of bST increased from about 8 ng/ml within one day to 222 ng/ml on average and declined thereafter to basal levels around day 17. Contrary to the tremendous increase in blood a measurable increase in milk was observed only on days 1-8. The maximum concentration in one individual sample was 4.2 ng/ml. IGF-I values are below 200 ng/ml before treatment and increased 5-7 fold thereafter. Control levels are reached around day 27 after the injection. IGF-I in milk is less than 10% of blood levels (about 16 mg/ml). After treatment there is an increase for 10 days up to 24-26 ng/ml with a decrease thereafter.

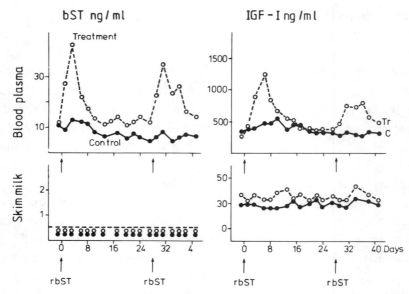

Figure 1. Concentrations of bST and IGF-I in blood plasma and skim milk of cows during treatment with a slow release device rbST (640 mg s.c.) or placebo (n=4 cows/group)

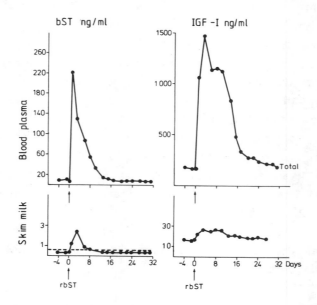

Figure 2. Concentrations (mean) of bST, and IGF-I in blood plasma and skim milk of cows (n=6) treated with a provocative high dose of slow release device rbST (1.92 g s.c.)

DISCUSSION

Literature reports to date are limited and indicating that only low levels of somatotropin can be measured in milk from both treated and untreated cows. Bourne et al. (1977) observed detectable bST in milk from untreated cows, but no increase was measured following subcutaneous injection of bST. Mohammed & Johnson (1985) have reported detectable bST in milk, but no increase in milk bST concentration in heat-stressed cattle treated with bST. Data from Hart et al. (1985) indicated that an increased incidence of milk serum samples with detectable bST resulted when cows were treated with bST.

Our data on bST in milk suggesting that bST is secreted into milk only at minimal concentrations below 0.5 ng/ml skim milk. Treatment of cows with a slow release device preparation does not increase the concentrations.
The results are in agreement with reports by Heeschen (1988) and Torkelson et al. 1989. When cows were administered a sustained-release formulation of recombinant bST (either 500 mg or 600 mg every two weeks), no increase in milk bST were detected relative to untreated cows. In the later study with an extreme sensitive assay (detection limit 0.3 ng/ml) milk somatotropin was shown to remain at a low level (fluctuating in general between 0.1 - 0.4 ng/ml) over a full lactation in both treated and untreated cows. The relative higher levels reported by Mohammed & Johnson (1985) and those reported later may be due to the fact that the previous study analyzed whole milk rather than milk serum.

There is clear evidence from the literature that IGF-I and other growth factors were found in milk of different species (see Koldovsky & Thornburg 1987). IGF-I concentrations are relatively high prepartum in human milk and decrease rapidly post partum (Baxter et al. 1984, Corps et al. 1988). The later found an increase again during full lactation about 6-8 weeks post-partum. Concentrations in human milk were about 30 ng/ml prepartum, about 14-18 ng during the first 2 days post-partum, 6-9 ng/ml thereafter and increased to 18 ng/ml again. Under physiological conditions similar observations were made in cow milk. Colostrum contained very substantial quantities of insulin, IGF-I and IGF-II (Malven et al. 1987). IGF-I levels decreased from about 150 ng/ml at parturition to about 25 ng/ml on days 4-6 post-partum. The results suggest that the transfer of insulin and IGF-I and IGF-II from blood into mammary secretions is similar to that of circulating immunoglobulins. Ronge & Blum (1988) found similar concentrations in cow milk during the first 6 days post-partum. During lactation and after treatment with bST only limited data for IGF-I are available. Torkelson et al. (1988) measured IGF-I concentrations in unexposed cows. Levels ranged from undetectable to 31 ng/ml; mean IGF-I was 2.5 ng/ml. During treatment with sustained release rbST a small, but significant increase was measured after the second and third injection. However, concentrations of IGF-I in milk of treated cows were within the range of values detected in milk from untreated cows. After daily s.c. injection of Jersey cows with bST for 7 days IGF-I levels increased at day 7 from 3.2 ng/ml to 11.7 ng/ml. Concentrations in blood increased from 113 ng/ml to 415 ng/ml (Prosser et al. 1988).

Baxter et al. 1984 found that fresh human milk contains immunoreactive IGF-I, of which a significant proportion is not protein bound at neutral pH.
The limited data for IGF-I in milk indicate that IGF-I occurs naturally in milk at much lower concentrations than in blood and that a clear measurable increase was only observed after administration of a provocative high dose of bST in a sustained-release formulation.

ACKNOWLEDGEMENTS
The technical assistance of Mrs. E. Kürzinger is gratefully acknowledged. We thank Dr. Raiti from the National Hormone and Pituitary Program, National Institute of Arthritis, Diabetes and Digestive and Kidney Diseases,

Baltimore, Maryland, USA for the generous gift of USDA-bGH-B_1 and USDA-BGH-I-1 and Prof. P. Gluckman, Auckland, New Zealand for supply of IGF-I antiserum. We thank Elanco, Eli Lilly Comp., Indianapolis, Indiana, USA for the support of this study.

REFERENCES

Baxter, R.C., Zaltsman, Z. and Turtle, J.R. 1984. Immunoreactive somatomedin-C/insulin-like growth factor I and its binding protein in human milk. J. Clin. Endocrinol. Metab. 58, 955-959.

Bourne, R.A., Tucker, H.A. and Convey, E.M. 1977. Serum growth hormone concentrations after growth hormone or thyrotropin releasing hormone in cows. J. Dairy Sci. 60, 1629-1635.

Corps, A.N., Brown, K.D., Carr, J. and Prosser, C.G. 1988. The insulin-like growth factor I content in human milk increases between early and full lactation. J. Clin. Endocrinol. Metab. 67, 25-29.

Hart, I.C., Bines, J.A., James, S. and Morant S.V. 1985. The effect of injecting or infusing low doses of bovine growth hormone on milk yield, milk composition, and the quantity of hormone in the milk serum of cows. Anim. Prod. 40, 243-250

Heeschen, W. 1988. BST in der Milch. In: "BST-Symposium", eds. Ellendorf, F., Farries, E., Oslage, H.J. Rohr, K. & Smidt D., Landbauforschung Völkenrode, Sonderheft 88, p. 220-226.

Koldovsky, O. and Thornburg, W. 1987. Hormones in milk. J. Pediatric Gastroenterology Nutrition 6, 172-196.

Malven, P.V., Head, H.H., Collier, R.J. and Buonomo, F.C. 1987. Periparturient changes in secretion and mammary uptake of insulin and in concentrations of insulin and insulin-like growth factors in milk of dairy cows. J. Dairy Sci. 70, 2254-2265.

Mohammed, M.E. and Johnson, H.D. 1985. Effect of growth hormone on milk yields and related physiological functions of Holstein cows exposed to heat stress. J. Dairy Sci. 68, 1123-1133.

Prosser, C.G., Fleet, I.R. and Corps, A.N. 1988. Increased secretion of insulin-like growth factor I into milk of cows treated with recombinant derived bovine growth hormone. J. Dairy Research in press

Ronge, H. and Blum, J.W. 1988. Somatomedin C and other hormones in dairy cows around parturition, in newborn calves and in milk. J. Animal Physiol. and Animal Nutrition 60, 168-176.

Schams, D. and Schmidt-Polex, B. 1978. Usefulness of radioimmunological determination of bovine prolactin in milk. Milchwissenschaft 33, 418-421.

Schams, D. and Karg, H. 1986. Hormones in milk. In: Endocrinology of the breast: basic and clinical aspects. Eds. Angeli, A., Bradlow, H.L. & Dogliotti, L., Annals New York Academy Sci., New York, 464, pp

75-86.
Torkelson, A.R., Lanza, G.M., Birmingham, B.K., Vicini, J.L., White, T.C., Dyer, S.E., Madsen, K.S. and Collier, R.J. 1988. Concentrations of insulin-like growth factor I (IGF-I) in bovine milk: Effect of herd, stage of lactation and sometribove, USAN (recombinant methionyl bovine somatotropin). J. Dairy Sci. 71, Suppl. 1, Abstract P 152.
Torkelson, A.R., Dwyer, K.A., Rogan, G.J. and Ryan, R.L. 1989. Radioimmunoassay of somatotropin from cows administered recombinant bovine somatotropin. Endocrinoly in press

EFFECTS OF ADMINISTRATION OF SOMATOTROPIN ON MEAT QUALITY IN RUMINANTS : A REVIEW

P. Allen[1] and W.J. Enright[2]

Teagasc, The Agriculture and Food Development Authority, Ireland. [1]National Food Centre, Dunsinea, Dublin 15.
[2]Grange Research Centre, Dunsany, Co. Meath.

ABSTRACT

The literature on the effects of administration of somatotropin (GH) on meat quality in ruminants is reviewed. Four studies on cattle, only two of which were in a published form, and a single report on lambs were found. In all studies GH administration improved growth rate without any deleterious effects on meat quality. A wide range of parameters have been examined including muscle composition, collagen concentration, solubility and synthesis rate, pH, juice extraction, weight loss on cooking, colour and pigmentation, shear force, compressibility and taste panel assessments. The only significant differences reported due to GH were a reduction in intramuscular fat concentration in one report on cattle, lower scores for tenderness and overall acceptability in another report on cattle, and a higher collagen synthesis rate in lambs which was matched by a higher non-collagen synthesis rate resulting in no change in collagen concentration. The main conclusion is that much more information is needed on the effects of GH administration on meat quality parameters, but the information available suggests that no serious problems exist.

INTRODUCTION

A number of studies have been done on the growth and efficiency enhancing effects of somatotropin (growth hormone; GH) in ruminants and these are reviewed elsewhere in the proceedings (Enright, 1989). Growth rate and feed efficiency are generally increased by administration of exogenous GH. Where changes in carcass composition have been noted they are in the direction of reduced carcass fatness and increased lean. Due to these beneficial effects on growth performance and carcass composition, somatotropin is a likely candidate to replace steroid hormones, now banned for use in the EC. To be acceptable to the meat trade and to consumers, however, they must be shown not to adversely affect meat eating quality. This paper will review the available evidence on this issue and indicate what further work is needed.

STUDIES ON THE EFFECT OF GH ON MEAT QUALITY

To our knowledge very few GH studies have reported on meat quality attributes (Table 1). Sejrsen et al. (personal communication) used nine pairs of monozygous twin heifers which were of dairy type. One twin of each pair was given daily injections of 100 ug GH/kg body weight (BW) starting at a BW of 180 kg and continuing for approximately 16 weeks (final BW 274 kg). Growth hormone improved growth rate, increased lean and decreased fat in the carcass. Measurements related to meat quality included chemical composition and collagen content of the longissimus dorsi muscle, pH, shear force, colour, pigmentation and taste panel scores of quality attributes. Adam et al.(1985) used 19 Belgian Blue-White heifers of about 16 mo of age (approx. 400 kg), ten of which were given daily injections of 50 ug GH/kg BW for 19 weeks. Growth rate was improved with no significant change in carcass composition. Fat content of the 7th rib section, pH, juice extraction, shear force (trapesius), compressibility (illiospinal) and collagen content (longissimus dorsi) were measured. Wagner et al. (pers. comm.) used 58 beef steers of about 336 kg BW to study the effects of GH and estradiol alone and in combination. Growth hormone - treated steers were given 960 mg GH every 14 days (204 ug/kgBW/d) throughout the 20 week trial. Growth hormone increased growth rate, reduced fat content and increased protein content of the carcass. Weight loss on cooking, shear force and taste panel scores were measured. Kirchgessner et al. (1987) reported on the effect of daily injections of 100 ug/kg BW of recombinant GH on the chemical composition of the semitendinosus muscle of female Fleckvieh veal calves.

A single report has been found on the effects of GH on muscle quality and composition in sheep (Pell and Bates, 1987). This study examined collagen and non-collagen turnover rates in five sets of twin Scottish half-bred x Suffolk lambs (ewes and wethers). Recombinant bovine GH was

TABLE 1   Summary of studies which have examined the effects of growth hormone on meat quality in ruminants.

Reference	Animal type	No./trt.	Daily dose (ug/kg)	Duration (wk)	Effect on: Growth	Carcass composition[1]
Adam et al. (1985)	Belgian heifers 16 mo.	10	50	19	Yes	No
Kirchgessner et al.(1987)	Fleckvieh veal calves	12	100	9	Yes	--
Pell and Bates (1987)	Scot. half-bred x Suff. lambs	5 (twins)	100	12	Yes	--
Sejrsen et al.(pers. comm.)	Dairy-type heifers 8 mo.	9 (twins)	100	16	Yes	Yes
Wagner et al. (pers. comm.)	Beef steers 336 kg.	14	204	20	Yes	Yes

[1]Increased lean and decreased fat.

TABLE 2   Effect of growth hormone on muscle composition.

Reference	Variable	Control	Treated	Sign.
Adam et al. (1985)[1]	% DM	29.3	28.9	
	% fat	26.2	25.5	
Kirchgessner et al. (1987)[2]	% DM	26.4	26.4	
	% fat	1.4	1.3	
	% N	3.8	3.8	
Sejrsen et al. (pers. comm.)[3]	% DM	25.2	25.1	
	% fat	1.8	1.4	***
	% N	3.6	3.6	

[1]7th rib section   [2]Semitendinosus   [3]L. dorsi
*** $P<.001$

administered at a rate of 100 ug/kg BW to one of each pair from 9 to 21 weeks of age at which time protein synthesis rates were determined. Growth hormone increased growth rate by 20% in these lambs.

Effect of GH on muscle composition in cattle

Three reports which examined the effects of GH on muscle composition in cattle are summarized in Table 2, although one of these (Adam et al., 1985) does not relate to true muscle composition since a complete section at the 7th rib was analysed. Only Sejrsen et al. (pers. comm.) found a significant effect of GH treatment, with a 20% reduction in intramuscular fat content. This was almost twice the percentage reduction in total carcass fat despite the fairly low level of intramuscular fat in the control animals. If it was confirmed that GH reduced intramuscular fat disproportionately, then concerns about GH adversely affecting eating quality could be raised. It is generally considered that a threshold level of intramuscular fat content exists below which eating quality is adversely affected. This threshold will vary between populations and individuals within populations but levels as low as 1.5% are likely to cause concern. Kirchgessner et al. (1987) also found a reduction in intramuscular fat content in veal calves. Although the reduction was as much as 12% it failed to reach significance.

Effects of GH on meat quality in cattle

Aspects of meat quality which have relevance for both the processed and fresh meat trade include water holding capacity (WHC), collagen content and colour. Three studies which have measured parameters relating to these attributes in cattle are summarized in Table 3. In no case was a significant difference reported, although the small differences in % soluble collagen (Sejrsen et al., pers. comm.), % collagen and juice extraction (Adam et al., 1985) and weight loss on cooking (Wagner et al., pers. comm.) that were observed were all in the unfavourable direction. PH, usually a reliable indicator of WHC, was identical in

TABLE 3   Effect of growth hormone on muscle quality.

Reference	Variable	Control	Treated
Adam et al. (1985)	% collagen	2.25	2.53
	pH	5.7	5.7
	% juice extraction	24.5	28.4
Sejrsen et al. (pers. comm.)	% collagen	0.23	0.22
	% soluble collagen	18.4	19.2
	pH	5.5	5.5
	Colour	37.9	37.4
	Pigmentation	102	103
Wagner et al. (pers. comm.)	% wt.loss on cooking	17.1	18.1

TABLE 4   Effect of growth hormone on eating quality.

Reference	Variable	Control	Treated	Sign.
Adam et al. (1985)	Shear force, kg	5.9	5.8	
	Compressibility	112	111	
Sejrsen et al. (pers. comm.)	Shear force, kg	7.9	8.6	
	Panel scores:			
	Colour	3.2	3.3	
	Taste	2.7	2.9	
	Tenderness	1.1	1.1	
	Juiciness	2.6	3.0	
	Overall acceptability	1.6	1.7	
Wagner et al. (pers. comm.)	Shear force, kg	2.5	2.3	
	Panel scores:			
	Tenderness	11.2	10.4	*
	Overall acceptability	11.0	10.2	*

*P<.05

control and treated animals in the two studies in which it was measured (Sejrsen et al., pers. comm.; Adam et al., 1985). Higher weight loss on cooking (Wagner et al.,pers. comm.) and higher juice extraction (Adam et al., 1985) found in GH-treated animals would be explained by a higher intramuscular moisture content. Indeed, dry matter of the longissimus dorsi was lower in the study of Sejrsen et al. (pers. comm.) although the difference was not significant. Wagner et al. (pers. comm.) did not determine muscle composition, but moisture content of the carcass was reduced by administration of GH.

Effect of GH on eating quality in cattle

Three studies have reported on tenderness and other quality attributes either determined mechanically or assessed by a panel (Table 4). Only in the study of Wagner et al. (pers. comm.) was a significant difference found, with samples from GH-treated animals having lower scores for tenderness and overall acceptability. Since tenderness is a major component in the appreciation of beef it is not surprising that scores for these two attributes showed a similar reduction. What is a little surprising though, is that a similar difference in Werner-Bratzler shear force was not found. A small increase in shear force was found by Sejrsen et al. (pers. comm.) but this was not significant and panel scores for five attributes including tenderness were all similar for treated and control samples.

Effect of GH on protein turnover rates in sheep

The effect of GH administration to lambs on collagen and non-collagen concentrations and synthesis rates in two muscles was reported by Pell and Bates (1987; Table 5). The weight of both the biceps femoris and the semitendinosus was significantly increased by GH administration which is not surprising given a 20% increase in growth. Non-collagen protein synthesis rate was increased ($P< .05$) by 30% for the biceps femoris but was not significantly increased for the semitendinosus. Non-collagen protein concentration was similar in treated and control lambs. Collagen synthesis was

TABLE 5. Effect of growth hormone on protein synthesis in lambs[1].

Muscle	Protein concentration (mg/g)				Protein synthesis (%/day)[2]			
	Non-collagen		Collagen		Non-collagen		Collagen	
	Control	GH	Control	GH	Control	GH	Control	GH
Biceps femoris	116	114	5.7	6.2	4.2	5.4*	.57	.74*
Semitendinosus	120	117	3.4	3.9	4.0	4.3	.74	.70

[1]Pell and Bates (1987)
[2]The percentage of the protein mass synthesized per day.
*P< .05

also significantly increased in the biceps femoris but not in the semitendinosus. The ratio of collagen to non-collagen protein in both muscles was unaffected by GH treatment indicating that GH treatment would not be expected to cause eating quality problems related to collagen content.

SUMMARY AND CONCLUSIONS

The first and most obvious conclusion to reach is that very little evidence is available on the effects of administration of GH on meat quality. A single report was found for sheep and only four, one of which reported only on muscle composition, were found for cattle. Only one of these (Wagner et al., pers. comm.) reported a significant, albeit small, adverse effect of GH on eating quality. Another (Sejrsen et al., pers. comm.) reported a significant reduction in intramuscular fat content due to GH which may be considered to be an adverse effect in animals with an already low level of intramuscular fat. The fact that any other small but non-significant differences found were in the adverse direction, namely, shear force, collagen solubility (Sejrsen et al., pers. comm.), collagen content, juice extraction (Adam et al., 1985) and weight loss during cooking (Wagner et al., pers. comm.) suggests a need for caution.

Of course, until dose rates and duration are established for optimum effect on growth rate and carcass composition, it will not be possible to say whether meat quality will be adversely affected by GH. In this context it is encouraging to note that the dose rates and treatment periods used were all successful in improving growth rate (though not all significantly altered carcass composition), yet no serious adverse effects on meat quality traits were observed. Nevertheless, none of these studies could be described as large scale trials and the trial with the largest number of animals per treatment (Wagner et al., pers. comm.) found a significant effect on taste panel scores for tenderness and overall acceptability.

ACKNOWLEDGEMENTS

The authors thank Mary O'Brien for her help in preparing the manuscript. We also thank Cathryn Geraghty and Charlie Godson for their assistance in preparing the slides used in the presentation.

REFERENCES

Adam, J.J., Fabry, J., Ettaib, A. and Deroanne, C. 1985. [Effect of exogenous bovine growth hormone upon Belgian white blue heifer's meat]. Recueil de Medecine Veterinaire, 161 (8/9): 655-633.

Enright, W.J. 1989. Effects of administration of somatotropin on growth, feed efficiency and carcass composition of ruminants: a review. In "Use of Somatotropin in Livestock Production". Accompanying chapter.

Kirchgessner, Von M., Roth, F.X., Schams, D. and Karg, H. 1987. Influence of exogenous growth hormone (GH) on performance and plasma GH concentrations of female veal calves. J. Anim. Physiol. Anim. Nutr. 58 (1-2): 50-59

Pell, J.M. and Bates, P.C. 1987. Collagen and non-collagen protein turnover in skeletal muscle of growth hormone -treated lambs. J. Endocr. 115: R1-R4.

Sejrsen, K., Foldager, J., Klastrup, S. and Bauman, D.E. 1986. Effect and mode of action of exogenous growth hormone on body tissue growth in heifers. 37th Ann. Mtg. EAAP 1 : 467 (Abstr.) and personal communication.

Wagner, J.F., Cain, T., Anderson, D.B., Johnson, P. and Mowrey, D. 1988. Effect of bovine growth hormone (bGH) and estradiol (E_2B) alone and in combination on beef steer growth performance, carcass and plasma constituents. J. Anim. Sci. 66 (Suppl. 1) : 283-284 (Abstr.) and Personal communication.

EFFECTS OF pST ON CARCASS COMPOSITION AND MEAT QUALITY

M. Henning*, E. Hüster,*, R.E. Ivy**, E. Kallweit,*
F. Ellendorff*

*Institut für Tierzucht und Tierverhalten (FAL),
3057 Neustadt 1-Mariensee, Federal Republic of Germany

**Pitman-Moore Inc., Terra Haute, Indiana, USA

ABSTRACT

Recombinant porcine growth hormone was applied to two different German breeds. Experimental period was 70 days from approx. 50 kg to 100 kg liveweight. Twenty-two female Pietrain and 22 female Large White (Deutsches Edelschwein) pigs were housed and fed in a performance testing station. Daily i.m. injections of rpGH were administered to 10 animals of each breed. No differences in growth rate occurred between treated and untreated pigs. After slaughter at 100 kg liveweight lean meat percentage was significantly higher in the treated pigs compared to control groups within breeds. This was also reflected by all fat measurements and retail cut weights as well as in anatomically dissected muscles. Quality measurements such as pH, colour reflectance, conductivity and water binding capacity did not alter significantly in the carcasses of treated animals when compared to control groups. Breed differences in these traits were clearly seen between Pietrain and Edelschwein.

In recent years recombinant porcine growth hormone (pST) as a compound affecting a number of traits has been applied to various pig breeds. For our study two German breeds were chosen which are different in performance. Twenty-two Pietrains and 22 Deutsches Edelschwein (Large White) female pigs were finished in a performance testing station according to the progeny testing protocol (MPA-Richtlinien, 1983). Finishing period was from 30 - 100 kg liveweight, experimental period from approx. 50 kg - 100 kg, e.g. 70 days. Five mg of pST were administered daily to 10 animals of each breed. Five Pietrains and 5 Large Whites were sham injected, six of each breed served as untreated controls. All animals were slaughtered in the Institute's slaughterhouse in Mariensee. Meat quality was assessed as follows: pH (30, 45 min. post mortem (p.m.)), conductivity (45, 60 min. p.m.), rigor (45 min. p.m.) and pH, conductivity, rigor and colour reflectance 24 hours p.m. Dissection into retail cuts followed determination of fat thickness (at various points along the carcass) and loin eye area. Selected muscles were weighed separately (M. semimembranosus, M. semitendinosus, M. Psoas major, M. long. dorsi, M. supra spinam). Samples from M. long. dorsi were analysed for water binding capacity, drip loss, water absorption, dry matter and fat content, and hydroxyproline.

When control and sham animals were not significantly different ($p<0.05$) subsequent use of the term "control" was valid for these two groups.

Growth and development for both the finishing and treatment period were not different between treated pigs and controls. Expected breed differences were found for most of the traits, but will not be reported here in detail.

After slaughter visual grading (E,I,II,III etc.) yielded a better carcass conformation for treated Large Whites when compared to controls. This corresponds well to a higher calculated lean content in pST animals (49.5% to 55.5%). Even for the Pietrain that were all graded in E, percent lean was significantly improved from 55.4 to 63% for control and treated animals respectively. A reduced fat thickness for both breeds (average fat depth in LW from 25 mm to 18 mm, in Pi from 19 mm to 14 mm) also indicated improved carcass quality in response to pST. Retail cuts had a tendency to higher weights, but some of the muscles (M. semimembranosus, M. semitendinosus, M. psoas major and M. supra spinam in LW and M. psoas major and M. long. dorsi in Pi) differed significantly in relative weight between treated and control groups.

Meat quality assessment was different as expected between the two breeds, however, alterations were not significant between the treated and control animals for quality criteria such as pH, conductivity and water binding capacity. Göfo-scores (colour) were slightly but not significantly higher in the control animals of the Pietrain when compared to the pST-treated group. Large White controls had even lower scores compared to treated animals. There was hardly any difference in dry matter in the M. long. dorsi sample between and within breeds. Fat content was significantly reduced for the treated vs. control groups in both breeds ($p<0.001$ in LW, $p<0.01$ in Pi). Although the relative amount of collagenous tissue seemed to be higher in treated LW's when compared to controls, the difference could not be confirmed statistically. The pST-induced muscle enlargement did not affect meat quality in LW's. The pH-values suggest that the observed increase in lean in Large White does not result in an accelerated glycogen depletion indicative of an increased stress susceptibility. The already inferior meat quality of all Pietrains did probably not allow to detect a further decline in quality criteria even though lean was increased by 7.6% in the pST-treated group.

REFERENCES

MPA-Richtlinien (1983) Stationsprüfung auf Mast- und
    Schlachtleistung und Fleischqualität beim Schwein.
    ADS Bonn

POTENTIAL FARM LEVEL AND DAIRY SECTOR IMPACT OF THE USE OF BOVINE SOMATOTROPIN (BST) IN THE FEDERAL REPUBLIC OF GERMANY

J. Zeddies and R. Doluschitz

"Institut für Landwirtschaftliche Betriebslehre"
University of Hohenheim
Schloss Osthof Süd, 7000 Stuttgart 70
FRG/WEST GERMANY

ABSTRACT

Bovine Somatotropin (BST) is a naturally occuring growth hormone produced by dairy cattle with a regulative function in milk production. In the early 1980s biotechnological production in large quantities became possible. The pharmacological approval of BST had not been given for application in the dairy industry at the time this paper was prepared.

The cost analysis shows that production costs per kg milk can be reduced by BST-application.

The results of the farm level analysis depend highly on the assumed scenario. It can be stated that an increase in farm income is not possible by applying BST due to a strict limitation of milk production caused by milk quotas allotted to individual farms and the lack of other production possibilities. Given the possibility of an expansion in production, in particular extension of the individual milk quota, an improvement in farm income was found when applying BST.

The sectoral extrapolations indicate a reduction in the number of dairy cows when the national milk quota remains unchanged. This volume reduction would extend into the beef market. Fodder production from grassland declines as well as fodder production on arable land. Thus, grain production is negligibly increased.

By rating the sectoral effects - in particular the changes in beef and grain production - by macro economic weights and comparing the result with the sectoral income loss in overall agriculture, a slightly negative balance can be shown.

In spite of scientific based information to consumers that milk from BST-treated cows is innocuous to human health, a reduction in the demand for dairy products cannot be ruled out - at least in the short run. This would cause additional reductions in milk quotas, price reductions for some dairy products, and would divert surpluses into other markets as well as bringing about a deterioration in quality according to consumer opinion.

Application of BST is expected to lead to additional structural effects in agriculture. However, these are generally overestimated.

In order to avoid distortions in competition within the EC, a decision either to permit or ban the use of BST should be made uniform within the EC.

INTRODUCTION

Bovine Somatotropin (BST) is a growth hormone naturally produced by dairy cattle and has a regulative function in milk production. In the early 1980s bio-technological production in large quantities became possible. The pharmacological approval of BST has not yet been given for application in the West German dairy industry.

The purpose of this study is to quantify cost effects and farm level reactions in the West German dairy industry in case the application of BST is approved. Furthermore, long term consequences for the agricultural sector are quantified assuming certain acceptance rates with special regard to the development of the herd structure.

PRODUCTION STRUCTURE IN THE DAIRY INDUSTRY

Return and income of farm operations in the EC are determined largely by the production of milk. In West Germany more than one quarter of the total agricultural sales (27.7 %), is directly derived from milk production (Figure 1). Furthermore, part of the earnings from beef production must be included in milk production.

Since the beginning of the 1950s the production of milk in the EC and West Germany as well has been characterized by a continuously upward trend in the milk volume delivered to the dairies (Table 1).

The major reason for this development was that until the mid 1970s there was an enormous increase in milk yield. Since then the dairy cow stock has also increased noticeably. As a result production represented 131 % of domestic consumption by the year 1983. In 1984, the European Community tried to cope with this surplus by introducing an EC-wide milk quota system. Since then the number of cows has decreased dramatically while milk yield has fluctuated.

Figures 2 and 3 provide a better understanding of the regional distribution of milk production. It shows the dairy cow concentration in the different counties of West Germany and the EC. Traditional milk producing regions such as parts of Schleswig-Holstein, northern Lower Saxony and the southern Alpine Regions grow the most forage. More than 50 % and up to about 95 % of the agricultural land is grassland in those areas.

Dairy farming in the EC and West Germany as well is characterized by a medium performance level (Table 2) and a comparatively unfavourable herd size structure. Table 3 shows that milk production in the EC as a whole is characterized by a high percentage of small average herd sizes and tremendous structural differences exist in the member countries. In 1985, almost 47 % of all dairy farms had only less than 9 cows, in Italy 75 %, in West Germany 40 %. On the other hand, 74 % of farms in the United Kingdom and

Figure 1: 1985/86 Agricultural Sales, Earnings in %

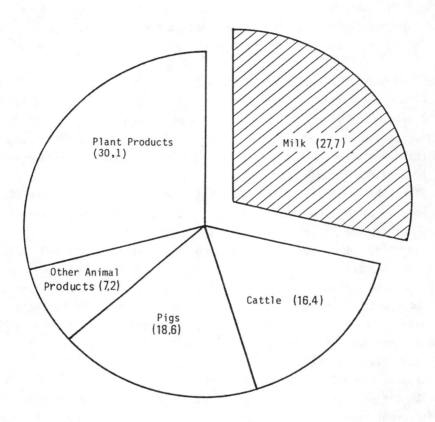

Source: Agrarbericht der Bundesregierung 1987

Table 1: Federal Republic of Germany: The most important milk production figures 1951 - 1987
(in 1,000 tons, if not otherwise noted)

	1951	1960	1970	1980	1981	1982	1983	1984	1985	1986	1987[1]
Dairy cow stock, Dec.-count (1,000 heads)[2]	5 804	5 800	5 561	5 469	5 438	5 532	5 735	5 581	5 451	5 390	5 074
Average milk yield cow/year (kg)	2 600	3 395	3 800	4 548	4 527	4 649	4 824	4 607	4 629	4 833	4 700
Cow milk production	15 171	19 248	21 856	24 779	24 858	25 465	26 913	26 151	25 674	26 350	24 450
fed to animals	1 618	1 869	1 425	1 010	986	973	982	1 104	1 282	1 385	1 475
Consumed by producers	1 735	1 513	1 063	580	538	527	493	473	460	472	
Processed by producers	715	732	319	36	38	37	36	39	40	40	766
Market milk sold ex farm	757	796	671	205	264	253	226	231	239	257	
Supplied to dairies	10 346	14 385	18 371	22 948	23 032	23 670	25 176	24 304	23 637	24 196	22 181
Supply in percent of production	66	75	84	93	93	93	94	93	92	92	91
Fat content of milk (%)	3,48	3,72	3,80	3,84	3,84	3,85	3,87	3,89	3,90	3,98	4,01
Producer price, free dairy (DM/kg)[3]	25,00	33,60	40,20	62,20	63,66	67,66	70,67	71,08	71,55	71,79	70,40
Low fat milk price (DM/100 kg)	4,70	4,80	5,70	9,20	10,45	10,62	10,28	10,87	11,48	12,20	
Low fat milk return	4 240	6 058	4 368	1 848	1 737	1 741	2 114	2 475	2 018	1 864	1 430
In percent of supply	41	42	24	8	8	7	8	10	9	8	6
Degree of self-sufficiency (%)	94	96	100	120	118	121	131	123	115	116	.

[1] preliminary
[2] since 1970 without nursing cows
[3] incl. VAT (value added tax)

Source: ZMP-Report 1986, Milk; Statistischer Monatsbericht BML 3/88

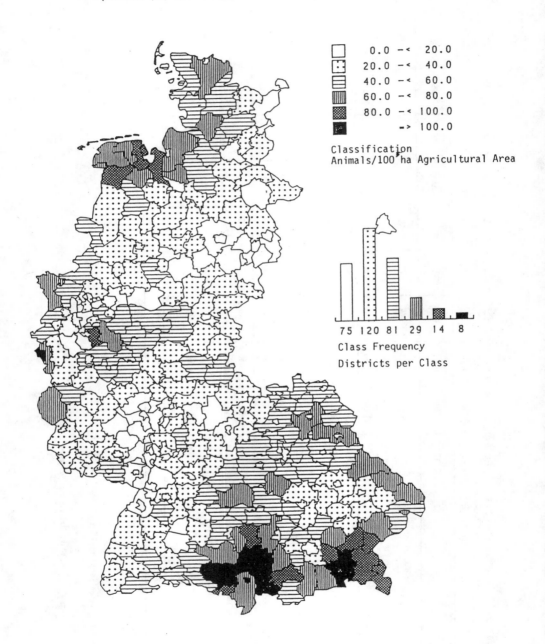

Figure 2: 1986 Dairy Cattle Population in West-Germany
(Animals per 100 ha Agricultural Area)

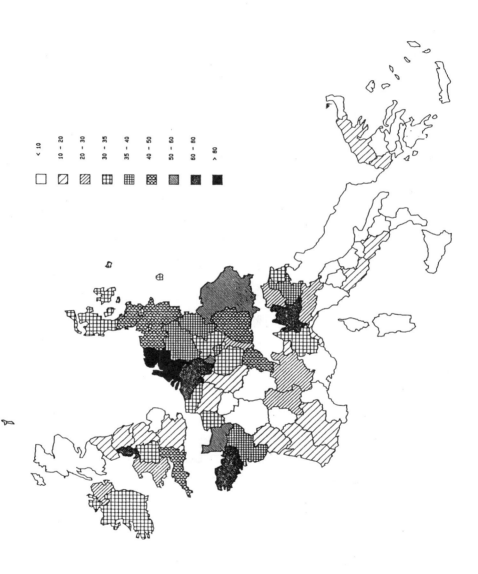

Figure 3: Dairy Cows per 100 ha Agricultural Land in the EC (10) in 1985

Table 2: Dairy Cow Production (1 000 each) and their Changes (%) in the EC Countries in 1970, 1980, 1983 and 1987

	1970	1980	1983	1987[1]	1980 vs. 1970	1983 vs. 1980	1987 vs. 1983
D	5 560	5 470	5 730	5 080	- 1,6	+ 4,7	-11,3
F	8 420	7 120	7 200	5 890	-15,4	+ 1,1	-18,2
I	3 560	3 010	3 220	3 060	-15,4	+ 7,0	- 5,0
NL	1 900	2 360	2 520	2 170	+24,2	+ 6,8	-13,9
B/L	1 060	1 050	1 060	980	- 0,9	+ 1,0	- 7,5
UK	3 240	3 280	3 430	3 040	+ 1,2	+ 4,6	-11,4
IRL	1 710	1 450	1 540	1 580	-15,2	+ 6,2	+ 2,6
DK	1 150	1 070	990	800	- 7,0	- 7,5	-19,2
EC (9)	26 600	24 810	25 690	22 600	- 6,7	+ 3,5	-12,0
GR	-	240	240	250	-	0	+ 4,2
EC (10)	-	25 050	25 930	22 850	-	+ 3,5	-11,9
SP	-	-	1 880	1 840	-	-	- 2,1
P	-	-	330	380	-	-	+15,2

1) Estimated

Source : EUROSTAT, FAO, OECD, ZMP, Stat. Bundesamt

Table 3: Dairy Cows and Dairy Cow Farmers According to Dairy Farm Sizes in the EC 1985

	Dairy Cows in Herd Sizes of ... (in %)			Dairy Cow Farmers with Herd Sizes of ... (in %)			Average Herd Sizes (Animals/Farm)
	1 - 9	10 - 29	30 a. over	1 - 9	10 - 29	30 and over	
B/L	5,5	38,6	55,9	25,5	47,1	27,4	22,2
DK	3,1	28,3	68,6	17,9	42,5	39,6	28,2
D	12,1	50,0	37,9	40,1	45,7	14,2	16,0
GR	73,4	18,4	8,2	95,5	3,8	0,7	3,0
F	7,0	44,6	48,4	29,2	48,4	22,4	19,8
IRL	7,6	31,9	60,5	39,8	36,7	23,5	19,9
I	26,6	33,4	40,0	74,9	18,9	6,2	9,1
NL	1,5	13,6	84,9	16,3	27,0	56,7	39,4
UK	0,4	5,5	94,1	8,7	17,4	73,9	61,6
EG(10)	10,0	34,3	55,8	46,6	34,6	18,8	17,8

Source: Statistisches Jahrbuch 1987, Statistischer Monatsbericht des BML 8/87, eigene Berechnungen

57 % of the farms in the Netherlands kept 30 dairy cows or more. Figure 4 shows the average herd sizes in dairy operations throughout the EC (10).

In the EC generally, the average milk yield per cow increased from 3 280 kg in 1970 up to 4 580 kg in 1987, representing a total increase of some 40 % or about 2.3 % per year (Table 4). After the introduction of the milk quota system, in 1987 milk production increased only slightly compared to 1983. The reason for the reduction in the short term performance is apparently the reduction in the milk quota which has been in the range of about 20 % since 1984. It is expected that in the future the rate of increase of milk yield per cow will be about 80 % of the amount before the introduction of the milk quota. Figure 5 shows the projected impact on the number of cows through 1995.

Regarding the milk quota system, the administrations in some EC-countries and in West Germany decided to adopt the system of individual farm-based quotas (Formula A). The penalty for exceeding the quota was 75 % before and 100 % after 1987. Thus, marginal income loss caused by exceeding the farm quota is about 0.32 DM per kg milk while the marginal income loss is 0.35 DM per kg milk when the farm quota is/was not realized (Table 5). Given these incentives, German dairy farmers are forced to adhere strictly to the quota.

Quota transfer between farms is possible but restricted to an amount of 1 to 2 % per year in West Germany, in other countries probably some more. The payment for rented milk quota is about 0.10 DM per kg milk in West Germany. Purchase prices for milk quota are much higher in UK and the Netherlands. Since 20 % of the transferred milk quota reverts to the federal government the rent actually paid is 25 % higher. The quota transfer is linked to a transfer of land in West Germany and the Netherlands. Thus, in West Germany for each unit of 4 000 kg milk quota 1 ha land has to be transferred either rented or sold. Table 6 shows that quota transfer under favourable conditions results in a positive income effect whereas under unfavourable conditions no additional income can be expected.

After introduction of individual farm quotas the income effect of an increase in milk yield fell significantly. This is so because the requirement of any performance increase caused by breeding progress or growth hormone is an appropriate adjustment of the production pattern within the farm. That means that for example alternative production activities have to be given to substitute the reduced income from milk production. Table 7 shows the potential income effect of different yield improvement in dairy

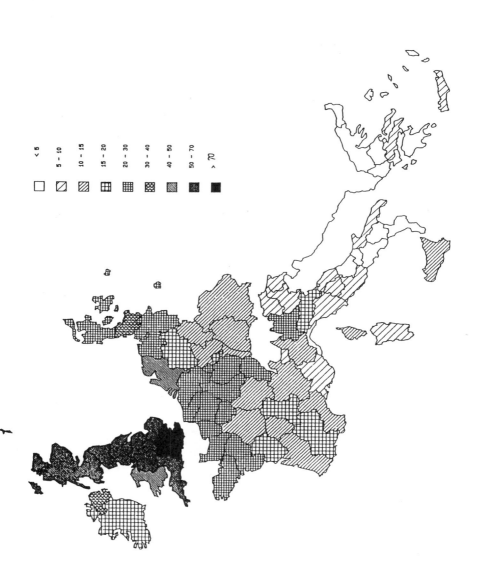

Figure 4: Average Herd Size in Dairy Operation in the EC (10) in 1985 (Cows per Operation)

Table 4: Average Milk Yield (kg/Cow and Year) and its Development in the EC

	1970	1980	1983	1984	1985	1986[1]	1987[2]
D	3 800	4 552	4 809	4 685	4 715	4 862	4 696
F	2 954	3 605	3 892	3 999	4 343	4 497	4 655
I	2 499	3 384	3 628	3 442	3 348	3 483	3 521
NL	4 300	5 030	5 181	4 947	5 203	5 439	5 389
B/L	3 512	3 859	3 953	3 869	3 943	3 968	3 909
UK	3 889	4 757	5 169	4 934	5 082	5 144	5 099
IRL	-	3 234	3 463	3 575	3 566	3 538	3 520
DK	3 884	4 846	5 411	5 504	5 685	5 916	6 065
EC (9)	3 280	4 185	4 499	4 297	4 458	4 591	4 580
GR	-	2 652	3 013	2 973	3 067	2 884	2 939
EC (10)	-	4 107	4 411	4 235	4 445	4 575	4 562
SP	-	-	3 325[3]	3 411[3]	3 363	3 373	3 406
P	-	-	2 490[3]	2 435[3]	3 189[3]	3 211[3]	3 042

[1] Preliminary Result
[2] Estimated
[3] Estimated by FAO

Sources: EUROSTAT, FAO, OECD, ZMP, Stat. Bundesamt

Figure 5: Development of Milk Production/Milk Quota, Average Milk Yield per Cow and Number of Cows in West-Germany

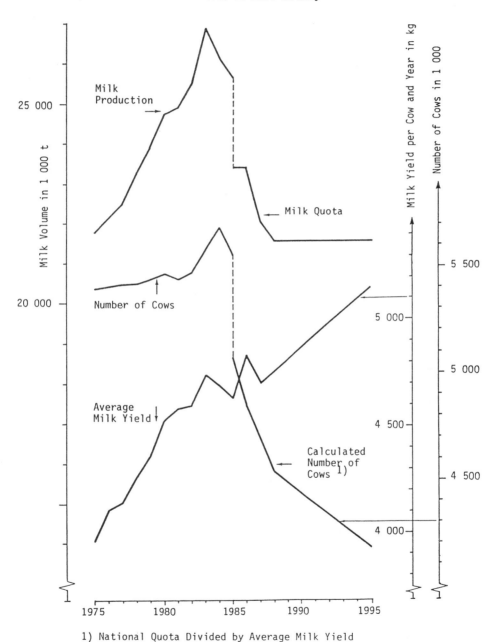

1) National Quota Divided by Average Milk Yield

Table 5: Income-effects of Over- and Under-fulfilment of Farm Quota (DM/kg Milk)

Year	1986/87	After 1987/88
Over-fulfilment		
Return	0,70	0,70
- Super duty[1]	0,50	0,67
- Marginal costs	0,35	0,35
Marginal income loss	0,15	0,32
Under-fulfilment		
Not realized return	0,70	
- Saved marginal costs	0,35	
Marginal income loss	0,35	

1) Until 31.3.87 75 %, after 1.4.87 100 % of the Reference Price

Source: RUTHS and ZEDDIES (1988)

Table 6: Income Effect of Dairy Production Considering Increase in Farm Quota

	Conditions for increase of farm quota			
	good (A)		not good (B)	
	DM/cow	DPf/kg milk	DM/cow	DPf/kg milk
Milk yield kg Milk-return (0,70 DM/kg) Return for calf and slaughter cow	5 000 3 500 747		4 000 2 800 747	
Total return	4 247	84,9	3 547	88,7
Variable cost Cost for roughage (0,4 DM/10 MJNEL)	1 353 942	27,1 18,8	1 005 969	25,1 24,2
Gross margin after subtraction of roughage cost	1 952	39,0	1 573	39,4
Payment for rented quota (0,10 DM/kg milk + 25%) Payment for rented land (1 ha/4000 kg milk) for 300 or 500 DM/ha respectiv.	652 375	12,5 7,5	500 500	12,5 12,5
Preliminary sum I	952	19,0	573	14,4
Gross margin from not needed rented land: (1,25 ha-0,5 ha) x 1000 DM/ha (1,00 ha-0,5 ha) x 500 DM/ha	750	15,0	250	6,3
Preliminary sum II	1 702	34,0	823	20,7
Income from alternative use of not needed labour capacity (60 h x 15 DM)	900	18,0	900	22,5
Income effect not considering investment	802	16,0	- 77	- 1,8
Yearly cost of investment in additional Buildings (5000 DM/Animal)	600	12,0	-	-
Income effect considering investment cost	202	4,0	-	-

Source: RUTHS and ZEDDIES (1988)

Table 7: Effects of Yield Improvement in Dairy Production

Situation		0	A	B	C	D
Milk yield per cow	kg	5 110	5 110	5 110	5 110	6 000
Fat	%	4,0	4,2	4,2	4,2	4,2
Protein	%	3,4	3,4	3,4	3,55	3,55
Number of cows		36,00	36,00	34,70	34,70	29,56
Farm quota [1]		183 960	177 337	177 337	177 337	177 337
Milk price	DPf/kg	70,50	72,36	72,36	73,50	73,50
Concentrates	dt/cow	11,4	12,4	12,4	12,9	19,9
Roughage	MJNEL/cow	23 009	22 807	22 807	22 651	21 090
Return	DM/farm	156 584	155 561	154 227	156 249	152 441
Variable cost	DM/farm	50 598	51 948	50 072	50 723	50 917
Gross margin	DM/farm	105 986	103 613	104 155	105 526	101 524
- Roughage	DM/farm	33 133	32 842	31 656	31 440	24 937
Difference I	DM/farm	72 853	70 771	72 499	74 086	76 587
- Savings in cost of not needed land[2]	DM	-	121	615	706	3 415
Difference II	DM/farm	72 853	70 892	73 114	74 792	80 002
- Savings in cost of not needed labour[3]	DM	-	-	1 170	1 170	5 796
Difference III	DM/farm	72 853	70 892	74 284	75 962	85 798

1) In A, B, C and D calculated delivered milk
2) DM/farm at 1000 DM/ha and 60 kg GJNEL/ha
3) DM/farm at 15 DM/h and 60 h/cow

Source: RUTHS and ZEDDIES (1988)

production. Comparing situation A with the basic situation 0 increasing the fat content reduces the individual farm quota accordingly due to the EC-regulation. In spite of an increase in milk price the returns will decrease whereas the variable costs of production will increase and the gross margin of the entire farm will be reduced. If the number of cows is adjusted as in situation B the gross margin will also decrease a little but using the released production factors (land and labour), the income contribution will be higher than in situation 0. Increasing the fat content will result in significant income gains if the released production factors land and labour are used productively. In comparison increasing the protein content or even increasing the milk yield (situation C and D) will not increase income.

FARM LEVEL IMPACT OF THE USE OF BST

Studies conducted in different research institutions (see Tables 8 and 9) show an increase in yield between 5.5 % and 30 % without supplying additional nutrients. These results are obtained based on the assumption that constituents remain unchanged. The cows under treatment showed an almost 10 % higher net energy intake. This means that the maintenance requirement, as well as the performance requirement per unit of milk is unchanged. The treatment would have no negative effect on the health and reproductive performance of the cow.

Other experiments are under way which indicate an increased feed conversion but on the other hand a prolonged calving interval as well.

Estimating the potential impact on the farm level parameters, in a first step, total cost of milk production with and without BST are calculated. The figures shown in Table 10 indicate that, excluding the cost of treatment, total production cost per kg milk could be reduced from 0.69 DM to 0.64 DM. Under this long term consideration an additional profit per cow per year of some 230 DM is possible. However, this calculation assumes a totally elastic adjustment of factor input according to changing performance capacity of the cows. This is unrealistic because the introduction of individual farm quotas under restricted possibilities of milk quota transfer inhibit or limit middle term adjustment of farm organization.

Five comprehensive country studies have been carried out since 1985: Berentsen et al. (1987), Buckwell and Morgan (1987), Lossouarn et al. (1987), Piva et al. (1987) and Zeddies and Doluschitz (1987).

Table 8: Influence of BST Application for several Months
(Daily Injection on Milk Yield)

Begin (Week of Lactation)	Duration (Days)	Dosage of BST (mg/d)	Yield of the Control Group (kg Milk[1]/d)	Increase in Yield by BST (kg/d)	(%)	Authors
3.	154	50[2]	19,9	4,0	19,9	PEEL et al. (1985)
13.	188	40,5[3]	27,9	11,5	41,2	BAUMAN et al.(1985)
13.	188	13,5[3]	26	8	30,8	HUTCHINSON et al.(1986)
13.	188	13,5[3]	No information		12,6	MOLLETT et al. (1986)
4.-5.	266	25[4]	25,7	4,7	18,3	BAIRD et al. (1986)
4.-5.	266	50[4]	24,2	8,9	36,8	CHALUPA et al. (1986)
4.-5.	266	25[4]	28,5	8,7	30,5	SODERHOLM et al.(1986)
4.-5.	266	25[4]	21,1	4,9	23,2	THOMAS et al. (1987)
4.-5.	266	50[4]	19,1	5,0	26,2	THOMAS et al. (1987)
4.-5.	266	25[4]	26,7	4,8	18,0	BURTON et al. (1987)
4.-5.	266	50[4]	27,7	5,4	19,5	CHALUPA et al.(1987

1) Corrected Milk at 3,5 % and 4 % Fat respectively
2) Only one dosage (p-BST, 0,78 IE/mg)
3) Max. effective Dosage (at 13,5, 27,0 and 40,5 mg r-BST/d)
4) Max. effective Dosage (at 12,5, 25,0 and 50,0 mg r-BST/d)

Table 9: Influence of long-term BST-compunds on Milk Yield
(Interval of Injection 14 and 28 Days respectively)

Begin (Week of Lactat.)	Duration (Days)	Dosage of BST	Yield of the Control Group (kg Milk[1]/d)	Increase in Yield by BST (kg/d)	(%)	Authors
6.-11.	168	960 mg/28 d[2]	21,8	1,2	5,5	FARRIES UND PROFITTLICH (1986)
5.-10.	168	960 mg/28 d[2]	25,3	3,1	12,0	ROHR et al. (1986)
14.	168	640 mg/28 d[2]	24,1	4,7	19,5	OLDENBROEK et al. (1987)
No inform.	84	960 mg/28 d[2]	21,6	3,1	14,3	MC GUFFEY et al. (1987a)
9.	224	No informat. (14 d-Interv.)	24,9	4,6	18,5	RIJPKEMA et al. (1987)

1) Milk with 4 % Fat
2) Max. effective Dosage 320, 640 and 960 mg r-BST/28d

Source: ROHR, K. (1987): Milchleistung und Milchzusammensetzung. Vortrag anläßlich des BST-Symposiums am 3./4. November 1987 in der Bundesforschungsanstalt für Landwirtschaft (FAL) in Braunschweig-Völkenrode.

Table 10: Total Cost of Milk Production[1] (DM/Cow and Year)

	Without BST	With BST[2] (+20% yield from the 100th day)
Milk Yield kg/Year	4 624	5 163
Roughage MJ NEL/Year	25 385	24 559
Concentrates dt/Year	8,0	11,3
Variable Cost	1 381	1 556
Cost of Roughage[3]	1 015	982
Cost of Buildings	800	800
Labour Costs	900	900
Subtraction of Additional Returns		
– Calf	400	400
– Slaughter Cow	525	525
Total Production Cost/Cow	3 171	3 313
Total Production Cost/kg Milk	0,69	0,64

1) On the Basis of the Yield Level of an Average Dairy Operation, Standard Income   30 - 50 000 DM
2) Without Cost of Treatment
3) 4 DPf/MJ NEL

Six farm models have been developed for the West German dairy sector (Table 11) and analysed by linear programming. Special emphasis has been put on animal nutrition, in particular on the yield depending part of nutrition. Two alternative assumptions were made regarding the impact of BST on milk production (BST use from the first and the one hundredth day of lactation respectively) and two different scenarios were considered to reflect potential future economic conditions for milk production in West German agriculture.

The results of the cost analysis are shown for one farm type as an example in Table 12 and Table 13. It can be stated that an increase in farm income is not possible by applying BST due to a strict limitation in milk production caused by milk quotas allotted to individual farms and the lack of other production possibilities. Given the possibility of an expansion of production, in particular extension of the individual milk quota, an improvement in farm income was found when applying BST (Table 14).

Table 15 presents the gross margins for all optimization runs of individual farms. The comparison of the alternatives 1 and 2 with the baseline situation (Alternative 0) shows that, due to the lack of utilization possibilities for released production capacities when using BST, the total profit contribution of the farm is negative in scenario A versus the baseline situation. In this situation the farm's adaption to the change in intensity of milk production must be achieved by reducing the herd size. Indeed it is possible to significantly decrease the production intensity on the grassland and raise grain production. However, all this does not compensate for the noticeable higher cost incurred for additional feed concentrates for milk production. Summarizing the calculation with the different alternatives for scenario A, it may be stated that under the condition of a strict limitation of milk production due to milk quotas allotted to individual farms and the lack of other production possibilities, a better farm operating result would only be achieveable in large mixed farms.

Assuming the possibility of additional leasing of milk quota for individual farms, the significance of a milk performance increase through hormone application gains in importance (Scenario B).

SHORT-TERM STRATEGY OF BST-USE

Due to the quota system the pre-conditions for BST-use are limited in

Table 11: Structure of Milk Production in West Germany by Farm Systems

Farm System	Delimination of Operations in 1 000 DM StOI[1]	Number of Farms 1000	%	Average Herd Size (Cows)	No. of Dairy Cows 1000	%	Average Milk Yield (kg/Cow)	Milk Production 1000 t	%
Forage-growing	<30	263,5[2]	35,4	6,2	1633,7	29,6	4054	6623,0	26,3
	30 - 50	63,6	8,5	21,4	1361,0	24,7	4624	6293,3	25,0
	>50	48,3	6,5	38,8	1874,0	34,0	5109	9574,3	38,1
Mixed	<30	40,0[3]	5,4	3,6	144,0	2,6	3899	561,5	2,2
	30 - 50	9,3	1,3	9,7	90,2	1,6	4530	408,6	1,6
	>50	8,7	1,2	14,7	127,9	2,3	5182	662,8	2,6
Sum Total/ Weighted Average		433,4	58,3	12,1	5230,8	94,8	4612	24123,5	95,8
Other Farms[4]		310,5	41,7	–	285,7	5,2	3684	1052,5	4,2
All Farms		743,9	100,0	–	5516,5	100,0	4564	25176,0	100,0

1) StOI = Standard Operating Income 2) Thereof 130 000 Part-time Farms
3) Thereof 20 000 Part-time Farms   4) Including Farms without Cows

Sources: - Agrarbericht der Bundesregierung 1985
- Statistisches Bundesamt: Land- und Forstwirtschaft, Fischerei. Fachserie 3,
  Reihe 2.1.4: Betriebssysteme und Standardbetriebseinkommen
  Reihe 2.1.5: Sozialökonomische Verhältnisse
- Eigene Berechnungen

Table 12: Farm Organization and Income[1] without Adjustment of Production (Scenario A)

		Without BST	With BST[2] (+20%, 100th day)
Dairy Cows	each	20,52	18,38
Heifers for Replacement	each	5,13	4,60
Heifers for Sale	each	4,10	3,68
Beef Bulls	each	9,23	8,27
Concentrates for Cows	dt	165,60	208,30
Grassland	ha	14,23	14,23[3]
Arable Land	ha	12,50	12,50
Thereof Forage-growing	ha	5,70	4,82
Grain Production	ha	6,80	7,68
Milk Quota	1 000 kg	94,90	94,90
Labour Input	Ak/Year	3 220	2 906
Labour Productivity	DM/Ak	19,74	21,47
Total Gross Margin	DM/Farm	63 547	62 406

1) Forage-growing Farm, StOI   30 - 50 000 DM
2) Without Cost of Treatment
3) Reduced Intensity

Table 13: Potential Income[1] from Alternative Use of Free Production Capacities by BST Use (DM/Farm)

Total Gross Margin with BST Total Gross Margin without BST	62 406 63 547
Difference I	− 1 141
Additional Income from Renting 1,55 ha Grassland to 300 DM/ha	+ 465
Difference II	− 676
Additional Income from Use of 314 Hours of Labour to 15 DM/ha	+ 4 710
Difference III − per Farm − per Cow	+ 4 034 + 219

1) Forage Growing Farm, StOI 30 000 - 50 000 DM

Table 14: Farm Organisation and Income[1] with Adjustment of Production by Increase of Quota (Scenario B)

		Without BST	With BST[2] (+20%, 100th day)
Dairy Cows	each	30,89	30,92
Heifers for Replacement	each	7,72	7,73
Heifers for Sale	each	6,18	6,18
Beef Bulls	each	0	0
Concentrates for Cows	dt	249,30	350,30
Grassland	ha	14,23	14,23
Arable Land thereof	ha	12,50	12,50
- Forage-growing	ha	9,93	9,40
- Grain Production	ha	2,57	3,10
Milk Production	1 000 kg	142,80	159,60
Labour Input	Ak/Year	4 101	4 101
Labour Productivity	DM/Ak	16,35	17,53
Total Gross Margin	DM/Farm	67 058	71 895

1) Forage-growing Fram, StOI 30 - 50 000 DM
2) Without Cost of Treatment

Table 15: Gross Margins Of Single Farm Optimization for all Farm Groups and Alternatives (DM/Farm)

Scenario	Farm Type and Size (1000 DM StOI[1])		Alternative 0	Alternative1 (+20%, 1st day)	Alternative2 (+20%, 100th day)
A	Forage Growing	< 30	18 689	17 356	18 122
		30 - 50	63 547	61 062	62 406
		> 50	120 274	113 887	115 786
	Mixed	< 30	30 112	29 351	29 736
		30 - 50	90 635	88 767	89 395
		> 50	173 926	174 442*	174 635*
B	Forage Growing	< 30	24 172	26 991*	26 451*
		30 - 50	67 058	74 887*	71 895*
		> 50	133 106	145 527*	139 832*
	Mixed	< 30	32 956	34 528*	34 179*
		30 - 50	93 291	97 267*	95 580*
		> 50	176 609	180 270*	178 526*

1) StOI = Standard Operating Income
* Income higher than in the Comparable Baseline Alternative (Alternative 0)

the EC. This is particularly true for an application throughout the whole lactation period. Therefore, short-term application strategies have to be evaluated. The following six alternatives have been considered in this evaluation:

Strategy 1: Beginning of BST-treatment dependent on the yield level of the cow
- 4 000 kg FCM: Start of application from the 57th day of lactation
- 5 000 kg FCM: Start of application from the 80th day of lactation
- 6 000 kg FCM: Start of application from the 100th day of lactation

Strategy 2: Treatment of those cows which show yield increases greater than average

Strategy 3: BST-application in certain lactation phases (i.e. only the first or the second, third and fourth quarter of lactation)

Strategy 4: BST-application in those periods of the year, where milk prices are high (for the FRG during summer)

Strategy 5: BST-application in yield-groups, i.e. treatment of cows with significantly low or average milk yield

Strategy 6: BST-application for short-term quota-adjustment.

In conducting an economic evaluation of those short-term application-strategies of BST it is important to consider that dairy farms have a fixed milk quota and thus the application of BST causes a reduction in the herd size. Therefore, production capacities of the farms fall free. Freed resources have to be used economically in other production alternatives if BST-use is to be profitable. The following calculations are conducted for a specialized dairy farm which can increase beef production. The reference system (BST-application throughout the whole lactation period) is shown in Table 16. No costs for BST and BST-treatment are considered in those calculations.

Depending on the calving time the increase in income per cow per year would be between 53 DM and 77 DM (Holstein-Frisian cows) and between 32 DM and 46 DM (Simmental cows) respectively. This calculation shows that there

Table 16: Gross Income from bST Application per Cow/Year in DM

Strategy	Calving time S(Spring) A(Autum)	Yield Kg FCM		Response in Kg FCM / Day +2	+4	+6
1. bST Application during whole Lactation (from the 100th Day post partum)	A	6 000	Holstein	84,18	77,50	203,84
	S	6 000	Frisian	44,93	53,34	106,24
	A	5 000	Simmen-	43,40	46,17	94,00
	S	5 000	taler	40,79	32,09	76,34
2. bST-effect at different Yield Levels	A	4 000		84,18	160,95	
		5 000		44,93	85,00	
		6 000		43,40	77,50	
		7 000		40,79	71,95	
3. bST-effect in different Lactation Phases: Quarter	2 A	6 000			8,40	
	3 A	6 000			22,37	63,25
	4 A	6 000			32,52	
4. bST in July, August, September assuming extreme price Fluctuation	A	6 000		103,12	187,72	230,68
	S	6 000		43,94	80,29	101,95
6. Quota Adjustment in case of missing the Target Quantity 0,56 DM/Kg FCM						

Source: BAUMBACH, 1988

are significant differences in productivity between cows of different breeds. Considering a Holstein-Frisian cow with calving time in autumn and a yield level of 6 000 kg FCM the expected income increase will yield at 77.5 DM per cow per year. The income increase will yield at 94 DM per cow per year if only cows with BST-response over average are treated (average yield increase per day per cow: 6 kg).

The income increase will be reduced when the average yield increase per day per cow is lower than average (2 kg per day per cow). More effective is the treatment of cows with extremely low yield levels. The BST-treatment can start earlier during the lactation period of such cows. On the other hand there is a reduced need to substitute concentrates for roughage which yields a higher expected income increase. The income increase shown for different quarters during the lactation shows that the efficiency of BST-application increases by the end of the lactation. This is particularly true for those farms which practice calving during autumn and apply BST during the last three months of lactation (July, August and September). Those farms would have an additional advantage from higher milk prices during summer time. Significant seasonal price differences are found in the FRG in the states of Schleswig-Holstein and Lower Saxony. In those states the milk price during summer is about 10 % higher than the average over the year.

If BST were used as a short-term adjustment to fulfill farm quota, an income increase of 0.56 DM/kg can be expected for a average dairy farm in the FRG. The marginal benefit per day of BST-treatment would be about 2.2 DM or 450 DM over a application time of 200 days (reference system) respectively. From strategy No. 6 (short-term quota adjustment) the highest income gain can be expected.

ASSESSMENT OF BST ACCEPTANCE FROM AN EXCLUSIVELY ECONOMIC VIEWPOINT

The application of BST is only acceptable on the individual farm, if the income difference after subtraction of treatment cost is positive. Under the assumption of a cost free application, it is evident, as is shown in Table 17, that potentially positive income effects due to BST can only be generally expected under the condition of an unlimited quota transfer. The positive income contribution might range from 125 to 250 DM per cow and year. The percentage of cows from the total number of cows in these operating groups is indicated in Table 17. Results from market studies for the Federal Republic of Germany show that BST will be applied

Table 17: Positive Gross Margin Differences of Alternatives with BST-Application versus the Baseline Alternative (DM/Cow and Year)

Scenario	Farm Type and Size (1000 DM StOI[1])		Alternative 1 (+20%, 1ST DAY)	Alternative 2 (+20%, 100TH DAY)	% Cows
A	Forage Growing	< 30	–	–	29,6
		30 - 50	–	–	24,7
		> 50	–	–	34,0
	Mixed	< 30	–	–	2,6
		30 - 50	–	–	1,6
		> 50	43,77	55,96	2,3
B	Forage Growing	< 30	197,27	158,81	29,6
		30 - 50	253,28	156,44	24,7
		> 50	248,17	134,47	34,0
	Mixed	< 30	184,94	143,88	2,6
		30 - 50	217,03	124,48	1,6
		> 50	245,54	128,66	2,3

1) StOI - STANDARD OPERATING INCOME

only in larger holdings, at least those with more than 10 cows per farm. Furthermore, it was estimated that if BST were approved for use in 1990 only 15 to 20 % of the larger farms would use the hormone. Under these assumptions about 21 000 dairy farms will accept BST and treat some 800 000 cows (Table 18). However, even this figure seems to be over-estimated since leasing of additional milk quota is an essential requirement but only about 2 % of the national milk quota is available for leasing or purchase per year.

ESTIMATE OF SECTORAL EFFECTS OF BST APPLICATION

To estimate the sectoral effects of BST application, reactions of individual farms are accumulated for the entire sector. However, it is essential to isolate the separate effect of structual change and the BST effect as it is shown in Table 19. The results of the sectoral extrapolation shown in Table 20 indicate a reduction in the number of dairy cows when the national milk quota remains unchanged. This volume reduction will extend into the beef market. Fodder production from grassland declines as well as fodder production on arable land. Thus, grain production is slightly increased. Even the sectoral total gross margin is decreased because 2.1 and 3.6 % of dairy farms respectively are forced to go out of business in order to release the required quota. The average gross margin per labourer remaining in the agricultural sector will increase. If released labourers achieve an average out of farm income of about 20 000 DM per year the total sectoral gross margin will be unchanged. From the results of this study it is apparent that application of BST will induce an additional effect on herd structure. As the graph in Figure 6 shows, the total milk supply increased from 1975 to 1983. Due to introduction of the national milk quota, production has to be reduced to less than 22 million tons in 1989.

After a temporary decline in average milk yield due to a significant reduction in the farm quota in 1984 and 1987, a further increase in milk yield must be expected in the future. In Figure 7 the development of performance of milk yield is shown under the assumption that it will reach 80 % of the average milk yield increase of the last five years before introduction of the milk quota regulation. At this point it should be realized that the breeding progress to be realized within the next 10 years is already genetically fixed in the cattle population. Figure 7 shows a significant predicted decline in the number of dairy cows in West Germany

Table 18: Assumptions on the Acceptance Rate of BST

Farm System and Size in 1 000 DM StOI[1]	No. of Farms (1 000)	Acceptance Rate (% Farms by Application in the whole Herd[2])	No. of Dairy Cows (1 000)
Forage-growing			
< 30	263,5	0	0
30 - 50	63,6	17,5 (11 130)	344,0
> 50	48,3	17,5 (8 453)	423,1
Mixed			
< 30	40,0	0	0
30 - 50	9,3	0	0
> 50	8,7	17,5 (1 523)	22,7
Total	-	(21 106)	789,8

1) StOI = Standard Operating Income
2) In Brackets: Absolute Number of Farms

Table 19: Expected Number of Dairy Operations with and without Application of BST in 1985[1]

Farm Type	Farm Size in 1000 DM StOI[2]	Initial Number of Farms (Alternative 0 in 1000)	Alternative 1			Alternative 2		
			No Acceptance (Farms)	Acceptance (Farms)	Total (Farms)	No Acceptance (Farms)	Acceptance (Farms)	Total (Farms)
Forage Growing	< 30	263500	224810	-	224810	232260	-	232260
	30 - 50	63600	43470	11130	54600	44870	11130	56000
	> 50	48300	39847	8453	48300	39847	8453	48300
Mixed	< 30	40000	40000	-	40000	40000	-	4000
	30 - 50	9300	9300	-	9300	9300	-	9300
	> 50	8700	7177	1523	8700	7177	1523	8700
Total	-	433400	364604	21106	385710	373454	21106	394560

1) Not taking into Account the structural Change independent of bSt-Application
2) StOI = Standard Operating Income

Table 20: Sectoral Effects[1] of BST-Application in West Germany's Milk Production

	Alt. 0	Alt. 1[6][7]			Alt. 2[6][7]		
		Caused by Changes in Herd Size	Caused by BST-Applic.	Total	Caused by Changes in Herd Size	Caused by BST-Applic.	Total
	abs.	abs. %	abs. %	abs. %	abs. %	abs. %	abs. %
Total Gross Margin in Million DM	18 335	-9,2 -0,05	-237,3 -1,29	-246,5 -1,34	-9,2 -0,05	-133,3 -0,73	-142,5 -0,78
Manpower (AK) in 1000 Heads	320	-6,8 -2,13	-11,6 -3,63	-18,4 -5,75	-6,8 -2,13	-6,8 -2,13	-13,6 -4,25
Total Gross.Margin/AK[2] in DM	57 296	1215 2,12	1387 2,24	2602 4,54	1215 2,12	819 1,43	2034 3,55
Dairy Cows[3] in 1000 Hds	5 218	-25,1 -0,48	-108,5 -2,08	-133,6 -2,56	-25,1 -0,48	-63,3 -1,21	-88,4 -1,69
Male Calves (Purchase) in 1000 Hds Swine in 1000 GV	1 454	-86,3 -5,94	0 0	-86,3 -5,94	-86,3 -5,94	0 0	-86,3 -5,94
Milk Volume[3] in 1000 t	24 036	0 0	0 0	0 0	0 0	0 0	0 0
Beef[4] in 1000 t LG Grain in 1000 t	2 484 / 10 745	-12,2 -0,49 / 129,1 1,20	-89,3 -3,60 / 339,6 3,16	-101,5 -4,09 / 468,7 4,36	-12,2 -0,49 / 129,1 1,20	-52,6 -2,12 / 204,2 1,90	-64,8 -2,61 / 333,3 3,10
Feed Production from Grassland in Million KStE	9 815	-345,0 -3,52	-411,0 -4,19	-756,0 -7,70	-345,0 -3,52	-348,0 -3,55	-693,0 -7,06
Concentrates for Dairy Cows[3] in 1000 t	3 996	2,8 0,07	391,3 9,79	394,1 9,86	2,8 0,07	259,7 6,50	262,5 6,57
Thereof Energy Rate III in 1000 t	0	0 0	553,0 -	553,0 -	0 0	0 0	0 0
Concentrates for young + fatten. Cattle in 1000 t	3 208	-85,7 -2,67	-117,1 -3,65	-202,8 -6,32	-85,7 -2,67	-69,0 -2,15	-154,7 -4,82
Mineral Fertilizer[5] - N in 1000 t	689	-13,7 -1,99	-12,2 -1,77	-25,9 -3,76	-13,7 -1,99	-8,1 -1,18	-21,7 -3,15
- P$_2$O$_5$ in 1000 t	581	-9,0 -1,55	-11,9 -2,05	-20,9 -3,60	-9,0 -1,55	-7,5 -1,29	-16,5 -2,84
- K$_2$O in 1000 t	787	-1,5 -0,19	-26,7 -3,39	-28,2 -3,58	-1,5 -0,19	-6,9 -0,88	-8,4 -1,07

1) Related to the Sector Balances of Forage-growing and mixed Farms
2) Actual required Manpower (AK) as shown in Optimum Solutions
3) Including Animals, as well as Production and Requirement Volumes from other Dairy Cow Operations
4) Of fattening Bulls, Heifers and old Cows
5) Taking into Account the nutritional Volumes from organic Farm Fertilizer
6) Including Production, Income and Costs on arable Land of Farms going out of Production
7) Including Beef Production, relating Income and Costs from Farms not taken into Consideration in this Calculation

Figure 6: Whole Milk Production

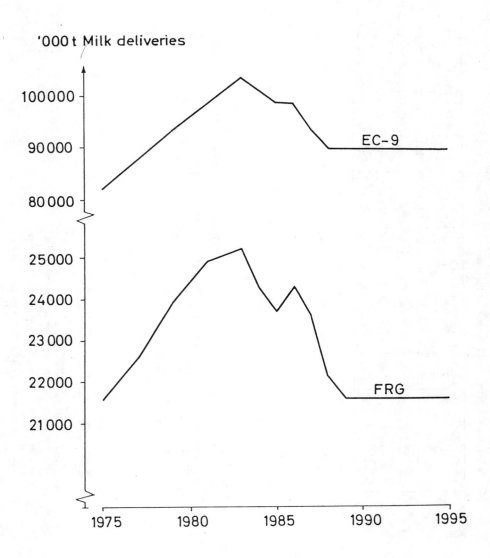

Figure 7: Structural Changes in Milk Production

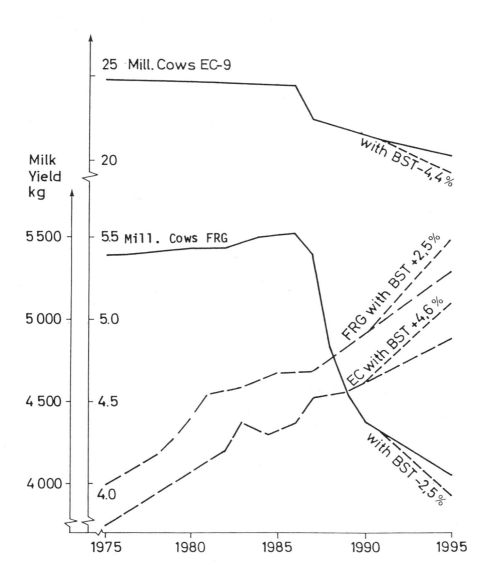

from about 5.4 million prior to 1985 to only 4.25 million dairy cows in 1990 and 3.73 or 3.95 million respectively in 1995, depending on whether BST is applied or not.

Nevertheless, it can be stated that the effect of BST on cow destocking in the range of 1.5 and 2.5 % respectively through 1995 is only a small portion of the necessary structural change which is caused by production adjustment to the demand. The average milk yield would additionally increase due to application of BST in the same relative order of magnitude.

MACRO-ECONOMIC IMPACT OF BST FOR WEST GERMANY

To derive the macro-economic effects of the application of BST, it is necessary to evaluate the effects generated in the different sectors from the macro-economic point of view. For marketable goods, the macro-economic value is determined by the price on the world market. Since the EC is generally taken to be the price leader for dairy products and beef in the world market, it follows that establishing macro-economic evaluation rates for marketable goods is extremely difficult. If, for the purpose of establishing an evaluation rate for milk, one takes the world market price for butter at the level of 35 % of the EC price, this results in a milk price of about 0.10 DM per kg on the basis of butter utilization. If the national economic value of milk protein per kg of milk is set at 0.07 DM following the substitution value for feeding purposes, this would result in a total price of 0.17 DM per kg of milk. Considering additional administrative costs and a milk price of about 0.75 DM per kg FCM, the difference of about 0.68 DM per kg FCM is interpreted as price support.

The export possibilities for beef on the world market are extremely limited in the medium term. This indicates that in the world market there is only a small share open because prices are politically influenced by government intervention to a high degree. In the past, the relation between world market price and EC price justified an evaluation rate of about 70 % of the EC price. This, however, has changed for the present and also for the medium term to something between 15 and 20 %. Therefore, price supports amounting to 6.00 DM per kg beef were assumed.

Basically, the macro-economic evaluation rate for grain may be derived from the level of world market prices. In the medium term it may be expected that the price support will come to about 250 DM per t, and the additional storage cost, due to the strongly increasing surpluses in grain, to another 150 DM per t which means in total, 400 DM per t of grain.

Changes in purchased feed-concentrates and mineral fertilizers become negligible from the macro-economic point of view.

Regarding the changes in number of employees in the agricultural sector, no special monetary evaluation has been carried out to consider detrimental impacts on already existing economic employment problems.

By weighting the sectoral effects with macro-economic prices (Table 21) and comparing the results with the sectoral income loss in overall agriculture, a slightly negative balance can be shown for West Germany. This figure can only give a very rough idea since it is just the national prosperity effect without considering the EC-internal financial transfer. Due to the common domestic market and common agricultural financing within the EC, financial transfers are taking place between the EC member states. In order to quantify the financial transfer, one has to consider the relation between share of financing and production.

IMPACT OF THE USE OF BST ON THE EC DAIRY SECTOR

The economic advantages offered by BST have been analysed for France, Germany, Italy, The Netherlands and the UK. The results were discussed by Buckwell (1987)[1]. The five studies using static LP models which produced a range of effects on farm gross margin are summarized in Table 22. All of these results were based on the assumption that the yield effect of BST is 20 %. These changes in farm gross margin do not include the cost of the BST. The negative results for Germany illustrate the situation where resources released from producing a given volume of milk with fewer cows means lower calf and cull cow revenues. The figures for Italy are influenced by higher milk price and it was assumed that BST not only increases yield but also improves the conversion efficiency of feed. The results for the other three countries show broadly similar ranges of effects. Given that the cost of the BST has to be derived from these figures it can be concluded that there will be many farm situations in which there is no net financial gain from using BST.

Concerning the rate of adoption of BST it is obvious that the target group of adopters is only a fraction of the current milk producers (Table 23). Buckwell (1987) assumed that this so-called "ceiling adoption level"

---

[1] Unpublished manuscript

Table 21: Evaluation of Sectoral Effects of BST-Application with Regard to National Economics

Parmameter	Changes in Production[1]		Value (DM/t)	Burden / Relief (-) (+)	
	Alternative 1	Alternat.2		Alternative 1	Alt.2
Total Gross Margin	-	-	-	-237,3	-133,3
Relief- Market Organization Cost					
- Milk	± 0	± 0	680	-	-
- Beef	-89,3[2]	-52,6[2]	6000[3]	321,5	189,4
- Grain	339,6[3]	204,2[4]	400	-135,8	-81,7
Balance	-	-	-	-51,6	-25,6

1) Exclusively caused by BST-Application
2) **Liveweight**
3) 6000 DM/t Slaughter Weight (60 % of **Liveweight**)
4) Including Production produced on arable Land from Farms going out of Production

Table 22: Improvement in Whole Farm Gross Margin per cow as a Result of BST treatment, 1986 prices

COUNTRY	Range of Improvement ECU/cow
West Germany	-158 to 181
France	4 to 114
Italy*	117 to 295
Netherlands	69 to 129
Great Britain	14 to 159

* The Italian study also assumed a 10% gain in feed efficiency

Table 23: Adoption of BST in the EC, 1995, by country

Country	Ceiling adoption level (Holdings > 30 Cows)	Proportion[1] of Herds	Proportion of cows
		percentage	
West Germany	12.9	5.5	18.4
France	20.0	8.5	23.1
Italy	4.2	1.8	23.3
Netherlands	59.2	25.0	45.5
Belgium	25.9	11.0	29.0
Luxembourg	46.7	19.8	37.5
United Kingdom	70.2	29.7	48.2
Eire	23.5	9.9	35.4
Denmark	39.1	16.5	38.6
EC-9	17.2	7.3	30.5

[1] Logistic adoption curve following the uptake of Artificial Insemination in UK (~ 40 % of potential farmers)

in each country is a current proportion of producers with over 30 cows per herd. This ranges from 4.2 % for Italy to 70.2 % for the UK. Based on the experience of the adoption of artifical insemination in the UK, a logistic curve was employed to calculate the adoption of BST which is shown in Table 22. The resulting adoption rate for 1995 for the EC (9) as a whole is calculated to be about 7 % of all herds making use of BST, accounting for 30 % of cows.

CONCLUDING REMARKS

In spite of scientific based information to consumers that milk from BST-treated cows is innocuous to human health, a reduction in the demand for dairy products cannot be ruled out - at least in the short run. This would cause additional reductions in milk quotas, price reductions for some dairy products, and would divert surpluses into other markets as well as bring about a deterioration in quality according to consumer opinion.

Application of BST is expected to lead to structural effects in agriculture. However, these are generally over-estimated.

In order to avoid distortions in competition within the EC a decision either to permit or ban the use of BST should be made uniform within the EC.

REFERENCES

Baumbach, U. 1988. Bovines Somatotropin (BST) in der Milcherzeugung und betriebswirtschaftliche Beurteilung eines gezielten selektiven Einsatzes. Diplomarbeit, Universität Hohenheim, West Germany.
Berentsen, P.B.M., G.W.L. Giesen and A.T. Oskam. 1987. The effect of using bovine somatotropin in the Netherlands. Landbouwuniversiteit Wageningen, The Netherlands.
Buckwell, A.E. and N. Morgan. 1987. Impact of Bovine Somatotropin in the United Kingdom in the context of Milk Quotas. Department of Agricultural Economics, Wye College, UK.
Lossouarn, J., F. Blanchard, C. Perrot. 1987. Somatotropin and the French milk industry. Centre d´Etude et de Recherche sur l´Economie et l´Organisation de Productions Animales, Paris, France.
Piva, G., A. Lazzari, D. Gattaneo and A. Farotto. 1987. BST - Economic Impact Study. University Piacenza, Italy.
Rohr, K. 1987. Milchleistung und Milchzusammensetzung. Vortrag anlässlich des BST-Symposiums am 3./4. November 1987 in der Bundesforschungsanstalt für Landwirtschaft (FAL) in Braunschweig-Völkenrode.
Ruths, F. and Zeddies, J. 1988. Milchquoten richtig managen. Referenzmenge besser einhalten. DLG-Mitteilungen H. 11/1988, p. 560-564.
Zeddies, J. and R. Doluschitz. 1987. BST Application and Potential Effects on Milk Production on Individual Farms and Sectors in the Federal Republic of Germany. Universität Hohenheim, Institut für landwirtschaftliche Betriebslehre, FRG.

APPLICATION OF BOVINE SOMATOTROPIN (BST) IN MILK PRODUCTION: POSSIBLE BENEFITS AND PROBLEMS

H. Glaeser and J. Gay

Commission of the European Communities
Directorate-General for Agriculture
Brussels

INTRODUCTION

By genetic engineering techniques, bovine somatotropin can be produced in any desired quantity. Applications for approval have been made and according to a recent IDF-paper, it is expected that BST will be officially approved before 1990 (Gravert, 1988). BST-application can increase the milk yield considerably. Published figures are in the range of 4 to 40%. There are several reports that the efficiency of milk production (kg milk/kg DM intake) was higher in BSTtreated animals. These data, which have to be critically evaluated, point to a considerable economic advantage of BST-application. On the other hand, there are a number of open questions, e.g.:
- Is there any influence of BST on milk composition, its technological properties and the quality of milk products?
- Will BST-application affect the structure of the EEC dairy sector?
- Which are the likely economic consequences for the EEC, if BST is applicated in other major milk-producing countries but not within the Member States?
- Is it possible to control BST-application?
- Is it possible to predict consumer reaction towards BST-application?

We shall try to discuss the above mentioned topics and to draw the attention to questions which make further investigations necessary. Animal health problems are beyond the scope of this paper because they are dealt with in another paper during this seminar.

EFFECT ON MILK YIELD

Published figures on the effect of BST on milk yield need a critical evaluation. For physiological reasons, BST-application is only useful from 10-15 weeks post partum. Furthermore, in practice only prolonged release preparations will be applied, which are injected every two to four weeks. Consequently, around 20% more milk can be expected per treatment period, which corresponds roughly to + 10% per year (Gravert, 1988). Some of the experiments carried out recently revealed, that there are considerable differences between cows in their response to BST-treatment. This topic will be dealt with more closely in the third section.

BST-treatment leads to a higher feed efficiency. In the above mentioned IDF-paper it is stated that "an increase of 20% milk corresponded roughly to 10% lower feed cost per kg milk produced".

EFFECT ON MILK COMPOSITION

Many reports clearly show that fat, protein and lactose levels in milk from BST-treated cows are not or only to a small extent changed, provided that the animals were adequately fed. Furthermore, Dutch and Italian studies seem to show that there is no significant influence of BST-treatment on technological properties (growth of starters, rennetability). These latter findings need confirmation.

On the other hand, it could be shown that the fatty acid composition of the milk fat was changed significantly. This is possibly a consequence of inadequate feeding or feed comsumption and not a BST-specific effect. Technological consequences with respect to buttermaking clearly can not be excluded. Recent German findings (Heeschen, 1988) led to the conclusion that BST-treatment can lead to an increase of the number of somatic cells and the pyruvatelevel in milk. The Council Directive on heat-treated milk (85/397/EEC) sets standards for cell counts which are lower than some of the results with milk from BST-treated cows. Furthermore, a "pyruric test" has to be carried out according to this Directive. A confirmation of these German findings would have negative economic consequences for BST-appliers. It should be mentioned, however, that other research groups were up to now unable to demonstrate a significant effect on the level of somatic cells.

In a recent IDF-paper (Teuber, 1988) Teuber pointed out that in several countries the effect of BST on milk composition (e.g. protein fractions, free fatty acids, enzymes) and technological properties is being investigated. Consequently, final conclusions cannot be drawn at present.

STRUCTURAL EFFECTS

There is some agreement that BST-application should not have major structural effects. The main arguments are:
- There is no need for expensive investments, the "small" farmer can, therefore, apply BST-treatment as well as the big one.
- BST can be applied in a "strategic" way, e.g. for exact quota adjustment, for the compensation of a production deficit caused by mastitis, for producing the same amount of milk with less cows.

- Economic analysis came to the conclusion that it is - on the basis of costs for the milk produced additionally by BST - as profitable for the small as for the big farmer to apply this technique.

Some of the arguments can be contested. For example, the economic advantage for an increased milk production using BST is dependent on the costs for the additional feedingstuffs which is required for that purpose. If concentrate is used, there may not be any advantage at all. The production of the same amount of milk with less cows is only feasible for farmers with 10 or 15 cows at minimum. In 1987, 44.5% of the farmers within the EC(10) had less that 10 cows totalling 9.0% of the dairy cow herd.

It can easily be calculated that it is impossible to produce the same amount of milk with 4 instead of 5 cows, if BST is applied. But even with a herd of 10 animals, this can be difficult or impossible, if for example only some of the cows respond in a sufficient way to BST-treatment. In practice, therefore, the economic advantage of producing the same amount of milk with less cows is only given for farmers whose herds have a certain minimum size.

ECONOMIC CONSEQUENCES OF BST-APPLICATION IN OTHER MILK-PRODUCING COUNTRIES BUT NOT WITHIN THE EEC

Prices for BST and costs of application are still unknown and at present it can not be said whether there will be an increase in animal health problems caused by BST. BST-application will probably be profitable for many farmers. Clearly, non-appliers will have an economic disadvantage. Consequently, the competitiveness of the EEC on the world market would be lowered, if BST is applied in major milk producing countries except EEC.

CONTROL OF BST-APPLICATION

For psychological (consumer reaction) or other reasons, one could share the point of view that there should not be any BST-application within the EEC. In this case, it should be born in mind, however, that there is until now no method for the control of BST-application. Consequently, one had to face the problem of illegal use of BST, which could be solved only by controlling production and trade. It seems necessary to state that at least in the past it was impossible to control the trade with pharmaceutical substances which were not admitted.

CONSUMER REACTIONS

The available information concerning this point is very scarce. There have been reactions by consumer associations expressing concern with respect

to a possible BST-application. An important question will be in which way
this topic is dealt with in mass media. It has to be explained to the public
that there is a considerable difference between artificial (synthetic)
steroid hormones, naturally occuring steroid hormones and protein hormones
like BST with respect to possible effects on human beings after ingestion.
Should this be impossible, there will probably be no chance to get public
acceptance for BST.

According to present knowledge, BST-application in milk production seems
to be safe:
- Being applied in normal amounts, there is no detectable carry-over into
  the milk produced.
- As a protein, BST will be digested after oral administration like other
  proteins.
- BST is species-specific, i.e. even after intraveneous application, it
  would not act as a growth hormone in human beings.

Theoretically, during digestion, peptides could be formed, which are
pharmacologically active (Heeschen, 1988). However, there is no evidence that
such substances are formed.

Summarizing this point, it can be concluded that there is presumably no
health risk, if BST is applied. However, it cannot be excluded that there
will be a reduction in the consumption of milk and milk products as a
consequence of negative consumer reactions.

There is evidence that consumers are increasingly concerned about animal
health/protection aspects with respect to meat production. It is possible,
therefore, that BST-treatment is also looked at by consumers from this point
of view.

CONCLUSIONS

The available information is not sufficient to draw final conclusions
with respect to BST-application. BST appears to be a powerful tool for
increasing the profitability of milk production; on the other hand, there are
questions which cannot be answered at present. Clearly, further research is
necessary in order to have a sound basis for decisions. If BST-application
does have economic benefits, and its possible drawbacks prove to be of minor
importance, there will be a strong pressure towards its application. A
renunciation of its use would cost tax payers' money. Clearly, this is a
speculative statement in the present situation.

Finally, I would like to draw the attention of the participants to the
statement of the European Parliament in the Happart Report (Happart, 1988)
and the resolution of 15.7.1988.

REFERENCES

Gravert, H.O. 1988. Current status of Somatotropin Research, IDF Document (Conference Room Document, Commission A, Working Group A 22), 72nd Annual Session of the IDF, Budapest 1988.

Happart, J. 1988. Projet de rapport sur les effects et les risques de l'utilisation des hormones de croissance et de la B.S.T. sur la production laitière et la viande, Parlement Européen (PE 115204).

Heeschen, W. 1988. BST und gesundheitliche Folgen, Schreiben an das Sekretariat der Enquete-Kommission "Technikfolgen - Abschätzung und Bewertung" des Deutschen Bundestages vom 6.5.1988.

Teuber, M. 1988. Influence of Bovine Growth Hormone on Composition and Quality of Milk, IDF-Document F Doc 161, 1988.

APPLICATION OF SOMATOTROPIN IN MEAT PRODUCTION:
IMPACT AT COMMISSION LEVEL

R. Nagel
Commission of the European Communities
Directorate-General for Agriculture
Brussels

INTRODUCTION

In the European Community, the meat sector represents around 30 percent of final agricultural production; it is the most important sector within the different agricultural activities and has an estimated value of 50 billion ECU each year.

Somatotropin is a growth-promoting agent which is of importance for the whole meat sector. Its possible use may influence considerably production techniques, economic results and perhaps also consumption pattern in the future. Therefore, a very careful examination of all the arguments in favour or against the use of this agent is absolutely necessary. Any decision on this matter concerns all the different meats; therefore, the final conclusions will be much more far-reaching than, for example, the ban on the use of sexual hormones which concerned more or less only steerbeef and veal production in some of the Member States.

CONSEQUENCES OF APPLYING BST IN LACTATION ON THE EEC-BEEF PRODUCTION
Meat from BST-treated cows

Research concerning the use of bovine somatotropin for milk production is well advanced and its practical use within the next few years seems, technically speaking, not excluded. It is therefore justified to examine the consequences of BST in milk production for the European beef market. Milk and beef production are closely linked together: 70 to 80% of our beef production has its origin in the Community dairy herd.

The first consequence, which could be regarded as less important but which should not be neglected totally, is the fact that beef coming from old dairy cows which have been treated with BST during the lactation, would arrive on the market. It may well be that this meat is totally free of BST residues and absolutely safe for the consumer. However, in principle, it has been "in contact" with BST. Therefore, any use of BST for milk production will automatically involve beef and the marketplace would be confronted with all the problems of meat quality, consumer safety and the reaction of the meat processing industry which uses this type of beef mainly.

May I add, for your information, that in 1987, 7,834,000 cows have been slaughtered in the Community, of which 75% are dairy cows.

Reduced number of cows leads to a lower beef production

The second consequence of the introduction of BST for milk production is of much greater importance and has already been the subject of different studies and estimations. The use of BST will increase the milk yield per cow. In view of the existence of the milk quota regime within the EEC, the global quantity of milk cannot be increased. Therefore, milk producers using BST must reduce their cow numbers if they want to respect their individual quota. In the beef sector, this has two effects: firstly, the use of BST will slightly increase in the short term the number of dairy cows taken out of milk production and ending up in the abattoir. Secondly, and in the more long term, less cows means less calves which results in lower beef production.

Forecasts in quantitative terms are very complex and difficult as long as we do not know the exact impact of BST on cow numbers. However, some figures have been published. A Study, undertaken by one of the pharmaceutical firms involved in research and production of somatotropin, predicts that with the use of BST beef production in 1993 would be reduced by 195,000 tonnes, compared to beef production without BST.

It is in this context it should be stated, that with or without BST, beef production will fall in the future, due to an annual increase in the milk yield as a result of improved feeding and housing techniques and genetic progress of the dairy herd. This will lead to a general fall in EEC beef production and hopefully allow us to reach a well balanced beef market. The use of BST for milk production will reinforce this process, but its effect will only be of a complementary nature to what is an already existing trend. Thiede (1987) foresees for the 8 main producing Member States (without Italy, Greece, Spain and Portugal) that in 1995 the total number of dairy cows - without BST - would reach 16.72 million head, 4.9 million less that in 1985. The use of BST in the EEC dairy herd would bring this figure to 16.26 million head, which means an additional reduction caused by BST of 460,000 head. It follows that the use of BST in the dairy herd would reinforce the fall of cow numbers which would normally occur by 10 procent.

However, having seen that reduction of cow numbers, one question has to be raised immediately: what will the farmers do with the land, the workforce, the houses etc, with the production capacities which will be "freed" at the farm level by the reduction of cow numbers. Already now, we see many farmers hit by the quota starting new activities, especially in beef. Many of the different agricultural branches for which no institutional barriers exist may be influenced by these "newcomers". So, in beef terms what we gain on one hand by the reduction of the dairy herd we may lose, partly or totally, on the other hand by more farmers going into beef.

EFFECTS OF USING SOMATOTROPIN AS A GROWTH PROMOTER

Somatotropin can also be used in the different species-specific forms, as a growth promoting agent for improving the fattening performance of cattle, pigs, sheep and poultry. However, research concerning the use of somatotropin for fattening animals is less advanced compared to BST in milk production and, in my opinion, many questions remain still open. It seems that there is a general handicap, at least for the moment, with regard to the use of somatotropin in the meat sector: the agent cannot be given to the animal by mixing it into the feed, but must be applied by injections or implants to each individual animal. This makes the application of the product in big production units very complicated and costly.

In general terms, the use of somatotropin for fattening results in increasing muscle mass and rate of gain and decreasing fat. In particular, the gain of lean meat and the reduction of fat mass may represent a positive element in the ongoing discussion on a healthy diet for the consumer. Although it is currently not possible to derive more than a "guesstimate" of the costs and benefits from the introduction of somatotropin, the feed cost reductions resulting from more efficient feed utilization would be expected to result in reduced costs.

Use of BST in beef production

Concerning the use of BST for beef, a lot of research was done for veal and steerbeef production which demonstrated an improvement of daily gain, an improved feed conversion efficiency and only a slight increase of feed intake per day.

The use of somatotropin increases the lean meat content and reduces the fat mass. This could be a very welcome effect for steer production, where the risk of a higher fat mass is much bigger than in the intensive production of young bulls. I am not aware of research and results concerning the use of BST for intensive young bull production. This is, however, in many Member States the main form of beef production. Without such results it will be difficult to evaluate the effect of BST on the future development of the European beef market.

Use of PST in pig production

Concerning porcine somatotropin (PST), research, especially in the United States, suggests the product has dramatic effects on pig growth rates and carcass composition.

We all know that pigs used in America and also in this experiment would be considered very fat by European standards. It is therefore understandable that research executed with European pigs shows the same effects, but of a lesser magnitude. It is worth noting that the lean meat content of the pig carcass plays an important role on the European market: producers are paid by the slaugtherhouses on the basis of the lean meat content of the carcass measured by objective instruments.

COMPARISON OF BST FOR MILK PRODUCTION WITH BST/PST FOR MEAT PRODUCTION

In this section, I would like to make a comparison, from the market division point of view, between BST for milk and somatotropin for meat production. For milk, we have a Community-wide quota regime which sets a global quantity of milk; the use of BST for milk cannot increase the global quantity and is, therefore, not a danger for the market equilibrium (under the condition that demand does not change). For meat the situation is quite different: there is no institutional restriction with regard to the quantities produced; the supply is not limited and the market is more or less free. The general use of growth promoting agents like somatotropin could lead to a significant increase of production which could jeopardize the balance of the different meat markets and weaken the farmers' income by lower market prices. With regard to beef, it is the intention of the Commission to propose a marked reduction of public intervention for the period beginning 1 January 1989; market forces will play a more important role in the future than today, and it will be the task of the producers to keep their production in line with demand.

CONSUMER REACTIONS

Indeed the word "demand" brings me to one of the key elements in the whole discussion, namely the attitude of the consumer vis-a-vis this new production technique, his acceptance of meat produced using growth promoting agents like somatotropin. Already today, meat is heavily criticized for being produced in an "unnatural" way, using legally or illegally all kinds of drugs, hormones and other chemicals. More and more, consumers ask for "natural" or "chemical-free" food products and do not trust scientific arguments which defend the wholesomeness of our agricultural products. The evaluation of the consumer's attitude is therefore a main element in the discussion on somatotropin, maybe not here, at our level, but certainly at the political level where the decisions will be taken.

INTERNATIONAL ASPECTS

Another element which should be taken into account when discussing the use of somatotropin in the EEC is our commercial relationship with third countries. We do not live on a well isolated island, where the policy can be established in total independence from what happens outside; we have an intensive import and export trade in meat, based in many cases on bilateral or multilateral agreements. To give you an idea of the size of this business, I would like to quote that the EEC imported in 1987:

500,000 tonnes of beef
100,000 tonnes of pigmeat and
260,000 tonnes of sheepmeat (without offals).

This meat comes from all over the world and is produced very often in countries where technical progress and new production techniques are readily used in animal production. As far as I know Sweden is the only country which banned the use of growth promoting agents, in 1986, for pigs; in many other countries, products like somatotropin may be allowed in the near future.

The trade in meat goes in both directions. The EEC is also an important meat exporter, for beef the biggest in the world. In 1987

910,000 tonnes of beef and
440,000 tonnes of pigmeat

were exported from the Community to third countries. For beef and pigmeat, export refunds can be granted to cover the gap between the EEC price level and the price level on the world market. It is evident that the use of somatotropin in those countries which are our competitors on the world market would widen this gap and could jeopardize our position on the world market.

I believe that it would be wrong to underestimate these trade problems. We should bear in mind that similar difficulties with third countries with regard to the EEC hormone ban are still unsettled.

FINAL REMARKS

I would like to end with two general remarks:

1. With regard to the use of somatotropin in meat production it is absolutely necessary to reach a single position for the whole Community. A situation where somatotropin would be allowed in some Member States but forbidden in other Member States would be a disaster for the common meat market, would lead to enormous distortions of competition and create new problems in intra-community trade. More controls at the border, higher costs and a lot of mistrust under the producers in the different countries would weaken the realization of the aim "1992".

2. With regard to any decision that might be taken on the use of BST for milk, it is important to bear in mind that this would be seen as a precedent for decisions concerning meat.

Different policies for two such related sectors would be difficult to defend.

REFERENCE

Thiede, G. 1987. Milcherzeugung im Somatotropin-zeitalter. Agra-Europe, 37-87, September 14, 1987.

THE EFFECTS OF TREATMENT OF DAIRY COWS OF DIFFERENT BREEDS IN A SECOND LACTATION WITH RECOMBINANTLY DERIVED BOVINE SOMATOTROPIN IN A SUSTAINED DELIVERY VEHICLE

J.K. Oldenbroek, G.J. Garssen
Research Institute for Animal Production "Schoonoord"
P.O. Box 501, 3700 AM Zeist (The Netherlands)

L.J. Jonker, J.I.D. Wilkinson
Elanco, Lilly Research Centre Ltd., Erl Wood Manor
Windlesham, Surrey (GU20 67 PH England)

INTRODUCTION

After calving 59 cows were assigned to a 252-day experiment to establish the effect of somidobove treatment on milk production efficiency, milk composition, metabolism, fertility and incidences of diseases in a successive second lactation. The cows were 11 Jerseys (group 1), 9 Dutch Red and Whites (group 2) and 13 Friesians (group 3), which had participated in a somidobove study in their previous lactation (Oldenbroek et al., 1988). Besides these three groups, 17 Friesians of parity >2 (group 4) and 9 Friesians of parity 2 (group 5) participated in this second-lactation trial. These cows had been treated with somidobove in their previous lactation for 12 weeks at pasture and for 12 consecutive weeks indoors.

From the 59 cows in this second-lactation trial 30 served as a non treated control group. They were not treated in the previous lactation or were treated with a low dose of somidobove (320 mg BST/28 days), which showed a limited response for the studied traits. The remaining 29 cows were treated in the previous lactation with a medium (640 mg BST/28 days) or with a high dose (960 mg BST/28 days) and they were treated in this second lactation trial from lactation week 13 onwards with the medium dose of somidobove.

In the 84 days preliminary period and in the 6 consecutive 28-day treatment periods the cows were grouphoused and individually fed ad libitum a complete diet of 50 per cent conc. and 50 per cent roughage which contained 62 per cent dry matter (dm), 6.78 MJ NE per kg dm and 126 g digestible crude protein per kg dm. Throughout the 252-day experimental period milk yield was recorded five days a week, milk composition two days a week, feed intake four days a week, feed composition biweekly and body weight once a week. On day 14 of all 9 28-day periods blood samples were taken to measure plasma levels of glucose, 3-hydroxybutryate, NEFA, urea, insulin, T4 and somatotropin. The same morning 4-quarter milk samples were collected to determine levels of calcium, magnesium, phosphorus and NEFA and to measure somatic cell count. Analysis of plasma samples for insulin and somatotropin has not yet been completed.

ANALYSIS OF DATA

The data was analysed with the analysis of variance model:

$$Y_{ijkl} = \mu + \alpha_i + \beta_j + \gamma_{k:j} + \xi_{ijkl}$$

where:

Y_{ijkl} = observation on animal l of group i, treatment j in second lactation and treatment k in previous lactation
μ = general mean
α_i = effect of group i, i = 1, 2, 3, 4, 5
β_j = effect of treatment j, j = 1, 2
$\gamma_{k:j}$ = effect of treatment k in previous lactation nested within treatment j in present lactation, k = 1, 2 : j=1 and k = 3, 4 : j=2
ξ_{ijkl} = error term

DIFFERENCES IN THE PRETREATMENT PERIOD (carry-over effects)

In Table 1 the average performance of 30 control (C) and 29 "treated" (T) cows in the first 12 weeks, the preliminary period, are presented.

From the averages in Table 1 and the curves in Figures 1-3 it can be concluded that in the preliminary period the "treated" cows showed a tendency to a higher feed intake than the control cows. No significant differences existed between the two groups in milk production traits and body weight. Nevertheless, significant differences were found for the efficiency of milk production and the energy balance. Animals, which were treated in the previous lactation with the medium or high dose mobilised less and stored more body tissue in the preliminary period than the control or low-dose cows of the previous lactation did. Especially in the first month of the preliminary period the latter two categories showed higher levels of 3-hydroxybutyrate and NEFA and lower levels of T4. In the previous lactation control and treated cows were fed the same diet and therefore control cows had more energy available for body tissue gain than the treated cows.

TABLE 1 Feed intake, milk production and body weight of control and "treated" cows in week 1-12 prior to treatment with somidobove and the effect of somidobove treatment in week 13-36.

	Week 1-12			Week 13-36		
	Control	"Treated"	P-value	Control	BST-effect	P-value
Net Energy Intake (MJ/day)	143	149	0.14	142	+3	0.32
Milk yield (kg/day)	32.8	32.3	0.67	23.8	+3.4	0.00
Fat yield (g/day)	1480	1441	0.51	1112	+182	0.00
Protein yield (g/day)	1043	1039	0.92	832	+107	0.00
Lactose yield (g/day)	1510	1478	0.60	1063	+153	0.00
Milk energy (MJ/day)	107	105	0.60	81	+12	0.00
Body weight (kg)	586	593	0.65	614	-10	0.12
Gross efficiency (%) *	75	71	0.04	57	+10	0.00
Energy balance (MJ/day)	1	9	0.04	25	-15	0.00

* Gross efficiency = milk energy : Net Energy intake

DIFFERENCES BETWEEN TREATED AND CONTROL COWS IN A SECOND LACTATION

In weeks 13-36 treated cows produced 3.4 kg milk, 182 g milk fat, 107 g milk protein and 153 g milk lactose more than control cows. In the first lactation these differences between the medium dose and the control were: 3.3 kg, 223 g, 130 g and 155 g respectively. In this second lactation no effect of treatment on fat percentage was found, while in the first lactation this was increased after the first two injections. Somidobove treatment significantly increased milk yield traits and milk energy production and had a significant negative effect on the energy balance. Therefore a substantial increase in the efficiency of milk production was found.

As in the first lactation somidobove treatment did not affect levels of calcium, magnesium, phosphorus and NEFA in milk. In the second lactation a tendency was found for a higher (46000 cells/ml) cell count in the treated cows during the somidobove treatment (Figure 4). This difference is not associated with differences in clinical mastitis between the groups and may only partly be associated with the higher milk yield of the treated cows. In the second lactation no significant differences between treated and control cows

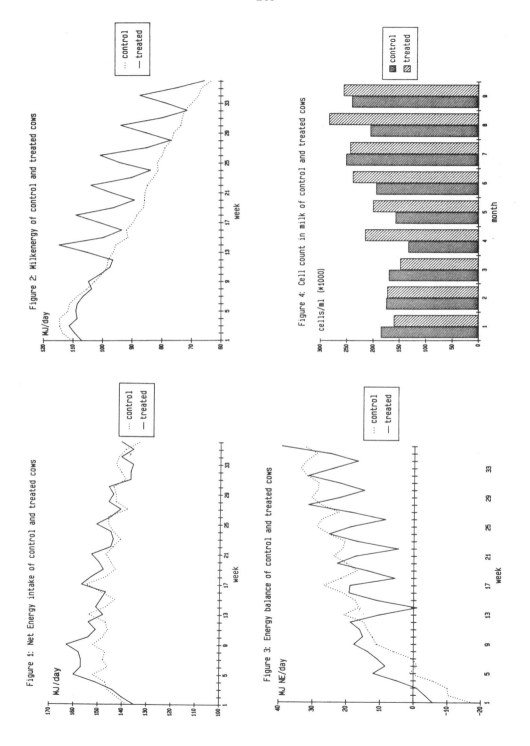

were found in plasma levels of glucose, 3-hydroxybutyrate and NEFA. Urea levels in plasma of treated cows were significantly lower than those of control cows. T4 levels tended to increase during the last months of treatment.

No significant differences between control cows and treated cows were detected for birth weight, gestation length, dystocia, retained placentas and inseminations per pregnancy. A tendency was found for a longer (1 week) calving interval in treated cows, partly associated with a longer interval between calving and first heat. Also a tendency was found for a higher incidence of twinning in treated cows. However, nearly all these twins had been conceived already before the start of the treatment in the respective lactation. In the second lactation no differences were found in the incidence of diseases between control and treated cows.

From the start of the first lactation up to the end of this second lactation experiment 10 out of 40 control cows had to be culled and 11 out of 40 treated cows. No differences could be found in the main reasons for culling: infertility, chronic mastitis and lameness.

CONCLUSIONS

Treatment of cows with somidobove in a previous lactation is associated with a slightly higher feed intake, a similar milk yield and therefore a lower efficiency of milk production in the first three months after calving. In this period treated cows had a higher energy balance than control cows.

Treatment with somidobove in a second lactation approximately gives similar increases in milk production and milk production efficiency as in first lactation.

More information on differences in metabolism, cell counts in milk, fertility including rate of twinning, incidence of diseases and reasons for culling will be obtained in the third and fourth lactation of this research project.

REFERENCE

Oldenbroek, J.K., Garssen, G.J., Forbes, A.B. and Jonker, L.J. 1988. The effect of treatment of dairy cows of different breeds with recombinantly derived bovine somatotropin in a sustained delivery vehicle. Contribution to the 38th Annual Meeting of the EAAP, Lisbon, Portugal. Accepted for publication in Livestock Production Science.

RECOMBINANT SOMATOTROPIN - A SURVEY ON A 2 YEARS EXPERIMENT WITH DAIRY COWS

P. Lebzien, K. Rohr, R. Daenicke and D. Schlünsen

Institute of Animal Nutrition and Institute of Production Engineering, Federal Agricultural Research Centre, Braunschweig-Völkenrode, Federal Republic of Germany

In the first year of our experiment, groups of 15 cows each served as controls or were injected with 320, 640 or 960 mg bST (SOMIDOBOVE, Eli Lilly) at 4 week intervals. During the first 12 weeks, animals were given fixed amounts of concentrates. Roughage was fed ad libitum. There were no differences in dry matter, net energy or crude protein intakes. During the last 12 weeks, feed allocation was according to yield, resulting in higher amounts of concentrates for treated animals.

Actual milk production and FCM production increased by 5 %, 8 % and 12 % with increasing bST-dosage (Table 1). There were no major differences with regard to milk protein and fat content.

TABLE 1 Dairy cow performance, first trial

bST-dosage (mg)	0	320	640	960
FCM (kg/d)	25.32^a	26.65^{ab}	27.29^{bc}	28.38^c
% protein	3.30	3.31	3.36	3.29
% fat	3.99	4.07	4.15	4.07

$a < b < c; \; p \leq 0.05$

Net energy efficiency (maintenance requirements plus milk energy plus net energy of weight change devided by net energy intake) was the same for control and treatment groups. A better feed conversion ratio of treated animals thus appears to be a dilution of maintenance requirements.

Each injection with slow-release bST vehicles caused a considerable yield increase with a sharp drop after two weeks. With the low and the medium dosage, milk yield approached that of the control prior to the next injection.

Individual cows differed in response to bST treatment. Part of the difference may be explained by differences in energy intake. For the time being, it is unclear whether actual milk yield or genotype will effect the degree of response.

In the second year, we used 14 cows which had been treated with bST in their previous lactation. 7 of these cows served now as controls, the other 7 were injected with 640 mg bST every 28 days. Another 7 cows - previously untreated - got also this dosage. As in the first year, forage consisted of a mixture of grass silage and maize silage (50:50), concentrate composition was also the same as in the first year. However, in contrast to the first year, concentrate allocation was according to energy requirements right from the beginning with forage being offered ad lib. Dry matter and net energy intakes of the treated animals were similar to those in the first experiment; intake of control animals were somewhat lower. bST-injections gave an increase in production of fat-corrected milk of 16 %. This effect was the same for previously treated or untreated animals. Differences in milk composition were minor and not significant. All animals were in positive energy balance. Similar to first experiment, there were no differences in net energy efficiency.

TABLE 2   Dairy cow performance, 2nd trial

	Control	640 mg bST (1st year)	640 mg bST (2nd year)
FCM (kg/d)	22.04^a	25.26^b	25.92^b
% protein	3.28	3.28	3.29
% fat	4.17	4.05	4.22

$a < b;\ p \leq 0.05$

In our experiments, we could not establish any negative health effects such as ketosis or other metabolic disorders. There were no differences in fertility between controls and treated animals. Results were generally poor with an average value of 2.6 services per conception and with calving intervals of 410 days in the first and 385 days in the second year. These poor results are partly due to the housing system which makes heat detection in the animals somewhat difficult. Reproduction will certainly be impaired with animals in substantial negative energy balance. Our recommendation would therefore be not to start bST injections before the third month of lactation.

EFFECT OF SLOW-RELEASED SOMATOTROPIN ON DAIRY COW PERFORMANCES

R. Vérité, H. Rulquin, Ph. Faverdin

Institut National de la Recherche Agronomique
Station de Recherches sur la Vache laitière
Saint-Gilles, 35590 L'HERMITAGE

Positive effect of somatotropin injections upon dairy cows on milk yield, has been broadly demonstrated by many experiments during the few last years due mainly to the development of the recombinant DNA technique for bovine somatotropin synthesis. However, many questions remain with regard to the variations in the responses of milk yield, level of intake and feed efficiency, to the incidence of (and the consequences on) the feeding techniques and to the side-effects on animal condition, health and reproduction on long term periods. Further, as daily injections are not feasible for practice, slow-released preparations have been developed for less frequent injections. However, their effects have still to be better characterized.

Three long-term studies have been conducted at INRA Station in Rennes with a sustained-released somatotropin preparation (somidobove supplied by Lilly Research Laboratories). The first trial (1985-86) was to determine the optimal dosage and to measure responses in the French situation i.e with diets based on good quality maize silage and rather moderate amount of concentrates. Trial 2 (1986-87) was to test the effect of somatotropin over a second lactation and to get information on the incidence of grazing vs winter diet. In the third trial (1987-88) two frequencies of administration (once every 2 or 4 weeks) were compared over two situations for protein nutrition (either medium or low-degradable protein delivered as normal or formaldehyde-treated soja and rapeseed meals).

The main experimental conditions are given in table 1. In each experiment, subcutaneous injections started 2 months at least after calving, following a preexperimental period. Ad libitum feeding of maize silage and appropriate concentrate allocations allow energy and protein requirements to be met on average but not necessarily for each animal or period, depending on the method of rationing (table 1).

In trial 1, milk, fat and protein yields showed a positive dose-response (11 %) up to 640 mg/injection but no further increase with 960 mg (table 2). In the low-dose group the increases were mediated through an increase of fat and protein contents but no such changes occured for the other groups. Those responses were almost constant throughout the experiment. Feed intake was slightly increased on average with the 640 mg-dose (+ 0.7 kg DM) but the differences appeared only during the second part of the trial when concentrate allocation was no longer equallized. Feed efficiency tended to be slightly increased but mainly during the first part of the experiment. Over the experiment, cows from the 640 and 960 mg groups tended to lose more weight and to recover lower bodyscore. Amount of body fat as measured by D2O dilution method (Vérité and Chilliard, unpublished) was decreased with 960 mg dose (- 15 kg) whereas it increased with control (+ 27 kg) and low-dose (+ 27.8 kg). The difference in body weight change arose mainly during the first half of the trial.

During trial 2 (table 3), milk was increased by 9 % during winter feeding but failed to change during the second phase at grazing, resulting in an average response of only 6 %. This lower response could perhaps be ascribed less to a negative long-term effect of repeated treatments over 2 years (as cows remained in good conditions) than to grazing itself, or to combined effect of pasture quality, frequency and lactation stages and season or to lower yielding ability over the second lactation (bad feeding conditions in early lactation affected negatively the potential yield of both groups). No significant effect appeared on body weight changes nor on feed efficiency, whereas feed intake during winter feeding was slightly increased (0,8 kg DM).

In the third trial (table 4), the quality of feed proteins did not affect cow performances at all, nor did it interact with the responses to somatotropin treatments, though PDI intake differed by some 400 g/day. However it should be noticed that animal requirements for intestinal digestible protein were always met even with the medium-degradable protein due to high level of intake (on average 21.4 kg DM of a 15 % CP diet). The means of somatotropin administration (either 320 mg every 2 weeks or 640 mg every 4 weeks) did not alter the responses. On average milk yield increased by 11 % and FCM by 12 %. However, protein content was decreased by 0.7 g/l. Feed intake was not changed on average, thus feed efficiency was increased by 10 %. Body weight gain was lower with the 4-weeks injection.

A common feature of these experiments was the cyclic pattern of the responses of milk yield and composition between two injections (figure 1). Milk yield and fat content increased rapidly over the first fews days after injections to reach a plateau between 5 and 10 days ; the effect vanished during the 3 rd week. In contrast, protein content sharply decreased (-1,5 g/l) within 24 hours of the injection ; this effect lasted for several days and recover in a happened during the 2 nd week. Blood NEFA and somatotropin level showed the same cyclic pattern as milk yield and fat, with maxima at day 10 whereas blood IGF1 was at maximum earlier (day 3) and decreased rapidly thereafter. Glucose level, when measured on day 1 (trial 2), exhibited a large transient increase. A decrease in urea level most often occured soon after injection, and disappeared by week 2. "Potential responses" in milk yield observed at maximum effect, differed largely from cow to cow in the range of 1 to 3 but appeared almost constant between successive injections within cow and experiment. They have to be questioned in relation to individual metabolic changes. Thus "potential responses" of fat content and NEFA level were directly related.

This work was supported by Lilly France S.A

Table 1 - CONDITIONS OF EXPERIMENTS

	Trial 1	Trial 2	Trial 3
Main treatments	4 doses BST	2 doses BST (2nd year)*	injections frequency (3) x protein degradability (2)
Side treatment (on 2 sucessive periods)	concentrate allocation	winter feeding vs grazing	/
Concentrate allocation			
. constant and equal for all cows	first 12 wks	-	16 weeks
. decreasing and according individual yields	last 12 wks	20 wks	-
Groups x cows	4 x 10	2 x 12	6 x 8
Duration (weeks)	6 x 4	5 x 4	4x 4 or 8 x 2

* Treated cows were taken from either of the 3 experimental groups, in trial 1, depending on calving date.

Table 2 -   EFFECT OF DIFFERENT DOSES OF SLOW-RELEASED SOMATOTROPIN ON COW PERFORMANCES (trial 1)

Dose mg/28 days		0	320	640	960
Milk	kg	24.79 a	+ 0.2 a	+ 2.69 b	+ 2.21 ab
Fat content	g p.1000	41.1 a	+ 2.3 a	+ 0.1 b	+ 0.4 a
Protein content	g p.1000	30.7 a	+ 2.1 b	+ 0.5 a	- 0.1 a
Concentrates	kg DM	4.89 a	+ 0.05 a	+ 0.35 ab	+ 0.77 b
Maize silage	kg DM	13.89	+ 0.39	+ 0.38	- 0.31
Body weight change	g/d	149	242	37	12
Body score change		0.57	0.52	0.32	0.17
Feed efficiency	kg FCM/FU	1.48	1.49	1.55	1.57
Energy balance	FU	0.6	0.8	0.2	0.1
Protein balance	g PDI	141	152	91	105

- Values for treatment effects are given as deviations from control (upper part) on as absolute figures (lower part).
- Within rows, means with different supercripts differ at $P < 0.05$).

Table 3 - EFFECT OF MONTHLY INJECTION OF SLOW-RELEASED SOMATOTROPIN OVER A SECOND LACTATION ON COWS PERFORMANCES (trial 2)

		WINTER FEEDING (8 first weeks)		GRAZING (8 last weeks)	
Dose	mg/28 days	0	640	0	640
Milk	kg	23.26 a	+ 1.99 b	18.62	+ 0.24
Fat content	g p.1000	39.7	+ 0.2	37.2	- 0.5
Protein content	g p.1000	29.3	- 0.3	30.2	+ 0.1
Concentrates	kg DM	4.84	+ 0.19	0.46	+ 0.27
Maize silage	kg MD	14.45 a	+ 0.84 b	/	/
Body weight change	g/d	58	144	308	398
Body score change		0.04	- 0.15	/	/
Feed efficiency	kg milk/FU	1.42	1.44	/	/
Energy balance	FU	0.79	0.80	/	/
Protein balance	g PDI	53	22	/	/

- Values for treatment effects are given as deviations from control (upper part) on as absolute figures (lower part).
- Within rows, means with different supercripts differ at $P < 0.05$.

Table 4 - EFFECTS OF FREQUENCY OF SOMATOTROPIN INJECTION AND PROTEIN NUTRITION ON COW PERFORMANCES (trial 3 - preliminary data)

		SOMATOTROPIN TREATMENT		
Frequency of injection		-	2 weeks	4 weeks
Dose	mg/injection	0	320	640
Milk	kg	26.9 a	+ 2.9 b	+ 2.8 b
Fat content	g p.1000	40.0 a	+ 1.7 b	+ 1.3 ab
Protein content	g p.1000	31.1 a	- 0.7 b	- 0.8 b
Concentrates	kg DM	6.20	+ 0.03	+ 0.01
Maize silage	kg DM	15.21 a	+ 0.29 a	- 0.21 a
Body weight change	g/d	172 a	131 ab	6 b
Body score change		0.12	0.01	- 0.07
Feed efficiency	kg milk/FU	1.40	1.55	1.55
Energy balance	FU	1.95	0.93	0.92
Protein balance	g PDI	247	138	151

- Values for treatment effects are given as deviations from control (upper part) on as absolute figures (lower part).
- Within rows, means different supercripts differ at $P < 0.05$.

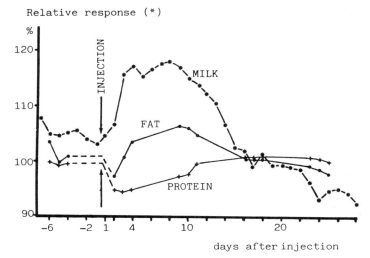

Figure 1 - Daily evolution of milk yield and composition following slow-released somatotropin injection (values from trial 2 are related to the means of days -7 à -1 and 22 to 28 over the 5 periods)

BST EFFECTS ON METABOLISM PARAMETERS IN DAIRY COWS
- Experimental Data -

E. Farries

Institut für Tierzucht und Tierverhalten (FAL),
3057 Neustadt 1-Mariensee, Federal Republic of Germany

ABSTRACT

The effect of rbST on some metabolism traits (glucose, ß-HB, GOT, GPT, urea, bST, IGF-1, fatty acids) was investigated in dairy cows over the period of 3 lactations. On the average, no significant difference could be found between treated and untreated animals. However, clear individual variations could be observed.

Feeding experiments with dairy cows of the breed "Deutsche Schwarzbunte" were carried out over a period of three lactations using various dosages of recombinant bovine somatotropin (rbST). We used a compound of the ELI LILLY Company in form of a slow release vehicle with a reaction time of 28 days. The hormone was applied subcutaneously.

The experiments were started between days 40 and 70 of lactation. In an additional experiment bST was injected directly after calving (days 4 - 7 post partum) for a period of 2 months in order to examine specific metabolic reactions in this highly catabolic period of lactation.

In total 3 experiments were carried out:

Experiment 1:
Control group	without rbST
Group L	= 320 mg rbST/28 days
Group M	= 640 mg rbST/28 days
Group H	= 960 mg rbST/28 days

Experiment 2:
Control group	without rbST in experiment 1 and 2
Group + -	= rbST in exp. 1, without rbST in exp. 2
Group - +	= without rbST in exp. 1, rbST in exp. 2
Group + +	= rbST in experiments 1 and 2

In this experiment we used the medium dosage of experiment 1 (640 mg/28 days)

Experiment 3:
Control group   = without rbST in all 3 experiments
Group + + +   = rbST in all 3 experiments
We used the medium dosage for experiment 1 (640 mg/28 days).

In an additional experiment with 6 dairy cows during early lactation we injected all cows with 640 mg/28 days.

The duration of experiments 1 to 3 was 6 x 28 days = 168 days each, experiment 4 lasted only for 2 injection periods, i.e. 56 days.

In addition to the daily registration of feed intake and milk yield, once a week (in experiment 4 twice a week) blood samples were taken from the udder vein to analyze glucose, ß-hydroxybutyrate, GOT, GPT and urea. Besides the contents of milk fat, milk protein and lactose, the concentration of urea and the composition of milk fat were determined simultaneously.

Results of the experiments:

The results for the metabolism data of experiments 1 and 2 show no significant differences between animals treated with various dosages and as compared to their controls. This is also the case for animals treated during two consecutive lactation periods.

If one looks for reactions in the 28 day periods, some tendencies become obvious: Values for glucose in bST-treated animals are sometimes more favourable which may indicate a certain support of the energy metabolism inspite of increasing milk yield.

On the other hand, ketone bodies are also somewhat increased underlining the lipolytical activity of bST. The concentration, however, does not exceed the physiological range. The intensified lipolytical activity is also reflected by the fatty acid composition of milk fat. Immediately after injection of rbST the fraction of long-chained, unsaturated fatty acids (C_{18}-$C_{18:3}$) increases.

With respect to the specific liver enzymes GOT and GPT there is no indication for a higher stress of the metabolism in treated animals.

In treated animals a tendency exists towards lower urea concentrations in blood as well as in whey. Own balance trials with dairy cows have shown a considerable drop in nitrogen excretion with the urine after rbST treatment. Thus, the organism seems to save up nitrogen for the higher milk protein synthesis.

However, all results show considerable differences between individual reactions looking at the data of single animals. Not all animals are able to compensate the higher stress by increasing milk yield and show clear problems in energy metabolism. Therefore, if rbST should be introduced into practical feeding systems, it should be used selectively.

Likewise rbST should not be injected during early lactation, i.e. directly after calving. At this time the endogenous synthesis of bST is very high, which means that lipolytical activity is also high causing the well known disorders in metabolism at the beginning of lactation. If this process is intensified by the application of rbST, it cannot be compensated by the animal. The ketone bodies reach unphysiological levels and the liver enzymes indicate a severe stress situation.

According to the milk yield it can be established that the onset of performance in the lactation following the former rbST treatment is absolutely normal and corresponds to the higher age of the cows. The endogenous bST synthesis, active only at the beginning of lactation is, therefore, not influenced by the former application of rbST.

The reaction on rbST is similar in all three investigated lactations, as far as mean values are concerned. The total lactational milk yield changes according to the number of lactation. However, individual deviations become also evident in milk performance which underlines the recommendation for a selective use of rbST under practical farming conditions.

EFFECT OF BOVINE SOMATOTROPIN ON MILK YIELD AND MILK COMPOSITION IN PERIPARTURIENT COWS EXPERIMENTALLY INFECTED WITH ESCHERICHIA COLI

C. BURVENICH[*], G. VANDEPUTTE-VAN MESSOM[*], E.ROETS[*], J.FABRY[**] and A-M MASSART-LEEN[*].

[*]Department of Veterinary Physiology, State University of Ghent-Belgium, Faculty of Veterinary Medicine, Casinoplein, 24, B-9000 Gent-België
[**]Station de Zoötechnie, Centre de Recherches Agronomiques, Gembloux, Belgique

SUMMARY

The objectives of the present study were to investigate the effect of a combination of local (polymyxine) and systemic (sulfonamides) antibiotic therapy with a short treatment of s.c. administration of BST (recombinant methionyl-bovine somatotropin) on milk and lactose yield in the uninflamed glands during acute experimentally induced Escherichia coli mastitis. Therefore, twelve recently calved cows of the East Flemish Red Pied breed and in their 2nd lactation, were used.

Very soon after calving, when the cows are very susceptible for coliform mastitis, animals were divided randomly into two groups of equal number : 1) a control mastitis group and 2) a BST mastitis group. Both groups were infused i.mam.with a Escherichia coli suspension (P4 : O32) after the morning milking in two homolateral quarters. Then, 24 hrs later, they all received a local (homolateral quarters; polymyxine) and systemic (sulfonamides) antibiotic treatment. Moreover, in the BST mastitis group, recombinant BST was administered on day 2 after mastitis induction and from then on, daily for 10 consecutive days.

After i.mam. infusion of E.coli, changes in MP as well as in milk composition in the infected glands were observed. There was a decrease in the production of lactose, α-lactalbumin, casein and fat. In the control group a decrease of milk and lactose production of 54% (with respect to preinjection values) was observed during the second half of the sham treatment. In the BST treated cows the loss of these productions amounted to 40%. The results with all other milk components were in line with those of lactose (see Table 1). Concomitant changes in the concentration of different electrolytes, lactose and cells in milk were observed. There was a decrease of the lactose, α-lactalbumin and potassium concentration and an increase in the concentration of sodium, chloride, serum albumin and polymorfonuclear cells. The normalisation of lactose concentration, which normally occurs in the control group, was faster during BST treatment. The results with all other milk components were in line with those of lactose. However, no differences could be detected between the total cell count in both groups.

In the uninflamed glands no local symptoms of inflammation were observed. Furthermore no indications of severe alteration of the blood milk barrier could be demonstrated. In the control group there was a significant decrease in MP during periods 0, 1, 2, 3, 4 and 5 in comparison to pre-injection period -1. In the BST group a decrease of MP during mastitis was observed in the same range. However, in contrast to the control group, MP returned to pre-injection values under influence of the BST treatment. The stimulating effect on milk yield in the BST group was more pronounced during the last 5 days of the 10-day treatment period but lasted until the end of the experiment (26th day). Qualitatively, the results with lactose, α-lactalbumin and all other milk components were comparable with those of MP. There was no loss in the BST group.

It thus seems that during E.coli mastitis in recently calved cows, short-term treatments with BST might have a beneficial effect upon the recovery of milk and lactose production in the non-infected quarters.

INTRODUCTION

Acute inflammation is the result of an interaction between host and pathogen and must be considered as one of the first body defence mechanisms during infection. It is generally accepted that most cases of coliform mastitis in the bovine take place during the periparturient period. The more severe clinical signs occur between calving and peak lactation and especially immediately post partum. These animals develop peracute mastitis which is characterized by a loss of milk production (MP) in both infected and non-infected glands (Hill et al., 1979).

It is well established that bovine somatotropin (BST, growth hormone) administered to healthy dairy cows increases MP. Positive responses have been observed at all stages of lactation in both positive and negative protein and energy balances. In early lactation (average 12 weeks postpartum) daily injections of BST (51.5 IU/day) for 10 days increased milk yield by 15% (Peel et al., 1983). The results obtained prior to peak lactation are lower (6%) than those in late lactation (Richard et al., 1985). Composition of milk is not affected by BST. The increase in yields of fat, protein and lactose run parallel with the increase in MP.

The objective of the present study was to investigate the effect of a combination of an antibiotic therapy with a short-term treatment of BST on MP and composition during experimentally induced E. coli mastitis in lactating cows 3 to 9 weeks after calving.

EXPERIMENTAL PROTOCOL

Seventeen cows of the EastFlemish Red Pied breed, in their 2nd lactation and yielding 12 to 22 l of milk per day, were used. For good adaptation, the cows were transferred to the stable of our laboratory just after calving and were kept in tie stalls. Cows were fed concentrates at 0800 and 1600 h daily at a rate calculated to meet requirements and they had free access to hay and water. They were milked twice daily at 0800 and 1900 h using a quarter milker. Quarter milk yields were recorded separately, and morning and evening milk was bulked in proportion to yield. All experiments started 3 to 9 weeks after calving.

Two groups of animals were distinguished: a control mastitis group of 7 animals and a BST mastitis group of 10 animals. After a pre-inoculation control period of 5 days, both groups were infected i.mam. after the morning milking on day 0. An E. coli suspension, containing 10^4 cfu of strain P4:O32, was injected in each gland of the left udder half. The right glands were used as control. Then, 24 hrs later, all animals received a systemic and a local antibiotic treatment (i.mam. polymyxin B, 500.000 IU, Pomesul, Bayer and i.v. trimethoprim (1g) +sulfadoxin (5 g) twice a day, Duoprim, Wellcome). In the BST group, recombinant BST from Monsanto was administered on day 2 and from then on daily at 9.00 h for 10 consecutive days by subcutaneous injection (40 mg/day, see Fig.1). In the control mastitis group, the vehicle was administered daily instead of the hormone.

Milk samples from bulk milk were taken twice a day at 0800 a.m. and 1900 h. during the pre-inoculation control period and from the 3rd to the 26 th day after i.mam. infection. These samples were mixed proportional to the respective yields and used for the determination of the different milk components. For each component its production per 24 h and per quarter was calculated. During the whole course of the experiment, 7 consecutive periods were distinghuished: period -1, period 0, period 1, period 2, period 3, period 4 and period 5. Each period lasted 5 days, with exception of the mastitis period (period 0) which included only the inoculation day and the day thereafter. Consequently, BST or sham treatment was established during period 1 and 2. On the day of mastitis and the day thereafter sampling occurred every 3 and 4 hrs respectively.

RESULTS

To study MP during experimentally induced mastitis, differences were made between inflamed and uninflamed glands as suggested earlier (Burvenich, 1983).

Effect on the inflamed gland

After i.mam. infusion of E.coli, changes in MP as well as in milk composition in the infected glands were observed as expected. There was a decrease in the production of lactose, α-lactalbumin, casein and fat. In the control group a decrease of milk and lactose production of 54% (with respect to preinjection values) was observed during the second half of the sham treatment. In contrast, in the BST treated cows the loss of these productions amounted to 40%. Qualitatively, the results with all other milk components such as α-lactalbumin, casein and fat are in line with those of lactose (see Table 1).

Mastitis is not only characterized by a decrease in the capacity of the mammary gland to synthetize different milk components. There is also an alteration of the blood milk barrier which is characterized by concomitant changes in the concentration of different electrolytes, lactose and cells in milk. There is a decrease of the lactose, α-lactalbumin and potassium concentration, and an increase in the concentration of sodium, chloride, serum albumin and polymorfonuclear cells (Burvenich, 1983). From the present study it is clear that the normalisation of lactose concentration, which normally occurs in the control group, is faster during BST treatment. The results with all other milk components are in line with those of lactose. In contrast with the afore-mentioned parameters no differences could be detected between the total cell count in both groups.

Effect on the uninflamed gland

In these glands no local symptoms of inflammation were observed. Furthermore no indications of severe alteration of the blood milk barrier could be demonstrated. Therefore the decrease of MP in these quarters could be ascribed to general illness of the cows. In the control group there was a significant decrease in MP during periods 0, 1, 2, 3, 4 and 5 in comparison to pre-injection period -1. This indicates that MP did not return to the pre-injection level. In the BST group a decrease of MP during mastitis was observed in the same range. However, in contrast to the control group, MP returned to pre-injection values under influence of the BST treatment. The stimulating effect on milk yield in the BST group was more pronounced during the last 5 days of the 10-day treatment period but lasted until the end of the experiment (26th day). Qualitatively, the results with lactose, α-lactalbumin and all other milk components were comparable with those of MP. There was no loss in the BST group (see Table 1).

CONCLUSIONS

It seems that under our experimental conditions of E.coli mastitis in postparturient cows, short-term treatments with BST might have beneficial effects upon the recovery of this disease. Furthermore BST seems to dispose of different pathways by which it influences milk secretion. This hormone is not able to bring the milk yield of the inflamed gland at its original level but it represents a net stimulatory drive on the synthesis of the different milk components. Moreover, BST influences the restoration of the blood milk barrier in a positive way. BST seems to protect the heterolateral gland from a significant loss in milk yield (due to general illness) which is characteristic for E.coli mastitis soon after parturition.

It is noteworthy that in this study, the cows only received a local and a systemic antibiotic treatment and that no other drugs were administered. The observed effects must therefore be considered as a consequence of a combination therapy. For further studies, one might also consider to combine BST treatment with anti-inflammatory drugs and/or to establish longer BST treatments.

REFERENCES

Burvenich, C. (1983) Mammary blood flow in conscious lactating goats in various physiological and pathological (mastitis) conditions. Ph.D. thesis. University of Ghent, Gent, Belgium.

Hill A.W., Shears, A.L. and Hibbit, K.G. (1979) The pathogenesis of experimental Escherichia coli mastitis in newly calved dairy cows. Res. Vet. Sci., 26, 97-101.

Peel, C.J., Fronk, T.J., Bauman, D.E. and Gorewit, R.C. (1983) Effect of exogenous growth hormone in early and late lactation on lactational performance of dairy cows. J.Dairy Sci., 66, 776-782.

Richard A.L., McCutcheon S.N. and Bauman D.E. (1985) Responses of dairy cows to exogenous bovine growth hormone administered during early lactation. J.Dairy Sci., 68, 2385-2389.

ACKNOWLEDGEMENTS

This study was supported by a grant from the Belgian Institute for Scientific Research in Industry and Agriculture (Instituut voor Wetenschappelijk Onderzoek in Nijverheid en Landbouw, Brussels, Belgium). Recombinant methionyl bovine somatotropin was kindly provided by Dr. C. Peel from Monsanto (Brussels, Belgium). The E.coli strain, originating from Dr. Hill (A.F.R.C., Compton, England), was kindly supplied by Dr.P.Rainard (I.N.R.A., Nouzilly, France).

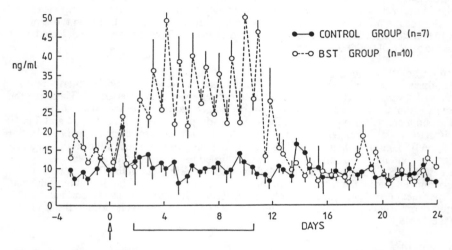

Fig. 1. Effect of sham or BST treatment (|_____|) of cows 3 weeks after calving on plasma BST during experimentally induced Escherichia coli mastitis on day 0 (↑).

TABLE 1 Mean percentage change in yields of milk, fat, casein, lactose and α-lactalbumin during the last 5 days of the 10 day BST or sham treatment period in periparturient cows suffering from E. coli mastitis. (- : loss; + : gain with respect to pre-infection values)

	Inflamed glands		Uninflamed glands	
	control	BST	control	BST
Milk production	-54	-40	-20	+2
Fat	-50	-37	-18	-1
Casein	-57	-31	-26	+8
Lactose	-54	-40	-23	-1
α-Lactalbumin	-61	-34	-43	-1

BOVINE SOMATOTROPIN - THE PRACTICAL WAY FORWARD RELATED TO ANIMAL WELFARE

W. Vandaele
Federation of European
Veterinarians in Industry and Research
Brussels, Belgium

INTRODUCTION

During the recent 4th European Conference on the Protection of Farm Animals, the Federation of European Veterinarians in Industry and Research (F.E.V.I.R.) had the opportunity to address the safety and benefits which the dairy farmer, the veterinarian and the dairy industry can expect from the first product developed by biotechnology for animals: the recombinant bovine somatotropin (Vandaele, 1988). Related to typical animal welfare implications, following points were stressed.

Animal health welfare implications are always studied. For instance, the opinions of the veterinarians involved in the trials are requested on animal welfare aspects.

Today trials are still continuing but from what has so far been published and from what we know already, we can conclude that dairy cows supplemented with BST just behave like high-yielding cows. In other words, if feed input is adapted to their yield, nothing abnormal happens.

High-yielding cows are already a fact in many countries or regions. In some herds or regions of The Netherlands, UK, and France, for instance, 7,000 and even 9,000 liter cows are already a reality with which dairy management has to cope.

One important point is the possible inter-relationship between high production and clinical diseases in dairy cattle: a recent review of 15 publications concluded that high milk production does not put a cow at increased risk of disease (Erb, 1987). Common diseases such as veterinary assisted dystocia, retained placenta, metritis, cystic ovary, ketosis, left displaced abomasum and mastitis are not more frequent in high-yielding than in low-yielding cows.

ANIMAL WELFARE IMPLICATIONS

Regulatory authorities can also ask expert groups to report on specific concerns.

For example, the UK Ministry of Agriculture turned to the Farm Animal Welfare Council (FAWC, 1987) for advice on the implications of using BST.

The FAWC concluded that there is no evidence of BST jeopardising treated animals' welfare in the short term. They added also that there were still "areas of uncertainty, which could have welfare implications where additional scientific informed opinion may be required" (FAWC, 1987).

In consideration of its findings, the Council recommends that guidelines for the commercial use of BST in the UK should be drawn up in collaboration with the Royal College of Veterinary Surgeons. The guidelines would include monitoring of herd performance and health with use of BST subject to veterinary supervision.

For confidentiality reasons, the companies cannot, at this time, provide detailed answers to all the concerns indicated in the FAWC report.

The registration dossiers submitted to regulatory authorities contain extensively safety data. Once this data has been rigorously evaluated by the regulatory experts, companies will be able to publish the evidence which meets these concerns. Let me, nevertheless, stress some general points:
- As earlier mentioned, BST is NOT injected before weeks 5-9 after calving when the cow returns to a positive energy phase.
- Long term reproductive efficiency is a classic parameter of safety.

In Europe there are no established regulatory guidelines for clinical trials with BST. But in the USA the Food and Drug Administration has indicated that data from two-year chronic toxicity tests are required plus one-year clinical trials in three different areas of the USA.

Responses to BST by cows with differing standards of energy and protein

Nutrition parameters are, of course, part of a clinical trial. From recent trials confirmation has been seen of the good response by cows.

In two trials (Monsanto), due to unfavourable climatic conditions, the grass was of bad quality. All the cows, treated and controlled, responded by reducing their milk output. The decline was greater in the BST group than in the control group, but the BST group still produced more milk than the untreated animals.

So, if farmers do not adapt the nutrient requirements, even supplemented cows simply do not respond.

The effect of BST on potential skeletal problems in heifers

This concern and aspects of performance of calves born from treated animals are answered in the registration data.

Feed back effect

There is no feed back effect after treatment with BST: when injections stop milk yield declines, but not below normal control levels.

Injection site reaction

One additional parameter for safety in the target species is evaluation of the potential reactions at the injection site mainly of course with slow release formulations.

As all trials are followed by veterinarians, they have to evaluate the injection site reactions and report if something abnormal occurs. Again, those data are part of any registration dossier.

Nevertheless, preliminary data have already been published (Whitaker et al. 1988, Phipps, 1988) related to two trials with a same 14 days slow release formulation to be injected subcutaneously. The authors concluded to be "satisfied that the fortnightly administration of BST by subcutaneous injection and the necessary handling was acceptable from the point of view of the animal welfare".

Also "there where no signs of distress, and subsequent any reluctance to enter the crush in which injections were administered".

It was also mentioned that the veterinarian had to look very carefully to observe the injection reactions and that farm staff did not even notice the swelling.

But the most important from the point of view of animal welfare, is that the cows showed no signs of distress since the animals behaved normally and were not reluctant to enter the crush in which the injections were made; also of course the milk yield response was maintained.

The concerns of FAWC on potential welfare abuses can best be answered by the fact that regulations for EEC countries foresee that BST will be used only under veterinary prescription.

REFERENCES

Erb, H., 1987. Can. Vet. Journal 28, 326-329.

FAWC, 1987. Farm animal Welfare Council report. Hansard (UK proceedings - House of Commons) dec. 1987.

Monsanto. Confidential information

Phipps, R., Weller, R.F., Austen, A.R., Craven, N. and Peel, C.J., 1988. Veterinary Record, pp. 512-513.

Vandaele, W., 1988. 4th European Conference on the Protection of Farm Animals, Brussels May 24-25.

Whitaker, D., Smith, E.J., Kelly, J.M and Hodgson-Jones, 1988. Veterinary Record, 503-505.

FIRST SIMULATION RESULTS ON THE IMPACT OF THE USE OF BST ON GENETIC GAINS FOR MILK YIELD

J.J. Colleau

Institut National de la Recherche Agronomique
Station de Génétique Quantitative et Appliquée
78350 Jouy-en-Josas, France

ABSTRACT

A very simple breeding scheme was generated by considering the progeny of 100 bulls (50 daughters per bull). The best 25 and 3 bulls were considered as sires of cows and bulls respectively. The best 1% of the daughters were considered as bull dams. The impact of BST was evaluated by comparing (%) the sum of the 3 selection differentials on the sire-cow, sire-bull, dam-bull paths with the corresponding value in the conventional situation, in order to get an approximation of what would become the annual genetic gain for milk yield in more complex schemes.

3 overall frequences of treatments were considered (10,30,50%) as well as 3 frequences of reporting of treatment (0%,50%,100%). 3 methods of allocation of BST were studied : at random, the best cows according to the normal yield without BST (supposed to be known), the worst cows on the same criterion. In addition, 2 very clear cut situations concerning the biological situation were analysed. In situation 1, BST brings forth to a mere translation effect of 1000kg per lactation, with no variation between cows ($r_G=r_P=1$). In situation 2, milk yield under BST is assumed to be relatively weakly correlated to normal yield ($r_G=r_P=0.8$).

The overall range of decrease of "genetic gain", over all the combinations, was 1-26% of the reference value (100 replicates per combination).

The major factors of variation were the genetic situation, (for which we do not know anything presently), with of course a detrimental effect of situation 2, and the system of distribution of BST. Except for a high frequency of treatment (50%), the use of BST on bad cows had the smallest impact. The largest impact were obtained for the use on the best cows. In that circumstance, a simplistic tentative to correct data with an additive model (without any prior information on the true parameters affecting the observed yield, and no consideration of possible untreated segments of lactation) brought more errors than if deleted (due to partial confounding between genetic levels and BST estimates).

Additional Monte-Carlo results are under way
1) in order to more accurately define the range of biological situations, starting with considerations on the supplementary yield allowed by the hormon
2) in order to measure the efficiency of data corrections based on multitrait BLUPs obtained from a mixture of normal (or so-called) total yields and early parts of lactation, before treatment.

VARIABILITY OF RESPONSIVENESS TO GROWTH HORMONE IN RUMINANTS: NUTRIENT INTERACTIONS.

J.M. Pell, M. Gill and D.E. Beever.

AFRC Institute for Grassland and Animal Production, Hurley, Maidenhead, Berks., SL6 5LR, UK.

Administration of exogenous growth hormone (GH) to ruminants has, to date, induced a consistent decrease in carcass fat content but the anabolic response, both in terms of liveweight gain and muscle growth, has been unpredictable. The aim of this investigation, therefore, was to determine the relationships between nutrient intake and hormonal status and their interactions in the control of growth.

Seventy-two lambs were randomly allocated to one of three protein diets containing 12 (L), 16 (M) or 20 (H) % crude protein (soya:fishmeal, 3:1) offered either ad libitum (A) or at 3% (R) of bodyweight (diet formulation, kg/tonne: barley, 740 (L), 645 (M), 545 (H); ground straw 170; protein 17 (L), 112 (M), 212 (H); molassine meal, 50; limestone, 23; mineral mix, 1). Within each diet, lambs were randomly allocated to daily injections of either saline (-) or GH (+, 0.1 mg/kg/d). All treatments commenced at 9 weeks of age and continued for 10 weeks; the animals were weighed and a jugular blood sample was taken weekly. At slaughter, visceral fat (mesenteric, omental, perirenal) was weighed and three skeletal muscles (gluteobiceps, vastus lateralis and semitendinosus) were dissected from the hindquarters. Results are shown in the Table; differences between groups were assessed by analysis of variance using initial liveweight as a covariate.

Treatment	Livewt. gain (g/d)	Empty body wt. (kg)	Muscle wt. (g)	Visceral fat wt. (g)	Total plasma IGF-1 at 10 weeks (ng/ml)
LR-	116.2	24.0	354.9	830	227
LR+	138.1	25.9	446.6	905	572
LA-	397.0	38.1	458.6	2258	509
LA+	385.0	41.8	590.5	2006	776
MR-	104.3	23.1	332.8	983	261
MR+	147.7	25.7	378.5	826	562
MA-	371.1	40.8	510.2	2409	666
MA+	389.3	40.4	512.1	1856	806
HR-	121.5	23.7	342.2	764	435
HR+	136.6	24.5	369.6	926	528
HA-	380.8	38.7	493.7	2078	711
HA+	404.1	44.3	572.0	1723	876
s.d.	33.6	3.3	65	379	158

GH induced a significant increase ($P<0.01$) in daily liveweight gain for restricted but not for ad lib. fed lambs. Food intake was slightly, though significantly ($P<0.05$), less in the GH-treated ad lib. than for their controls (5.5 versus 5.2% of liveweight). Overall, GH treatment resulted in a significant increase in combined muscle weight for both restricted and ad lib. fed animals ($P<0.001$), whereas visceral fat depots were significantly decreased ($P<0.05$) only in ad lib. fed lambs. IGF-1 concentrations were significantly increased by GH treatment ($P<0.001$), dietary energy intake ($P<0.001$) and dietary protein level ($P<0.05$).

The differential effects of GH on skeletal muscle and visceral fat do, in part, explain the variability in responsiveness to GH treatment in terms of liveweight gain. Further interpretation awaits a more detailed carcass analysis. Clearly, the interaction between GH status and dietary protein and energy levels is complex and requires further elucidation.

EFFECT OF RECOMBINANT PORCINE SOMATOTROPIN (rpST) ON FATTENING PERFORMANCE AND MEAT QUALITY OF THREE GENOTYPES OF PIGS SLAUGHTERED AT 100 AND 140 KG

P. van der Wal, E. Kanis, W. van der Hel, J. Huisman and M.W.A. Verstegen.

Wageningen Agricultural University, P.O. Box 338, 6700 AH Wageningen and TNO-Institute of Animal Nutrition and Physiology (IMGB, dept. ILOB), P.O. Box 15, 6700 AA Wageningen.

ABSTRACT

Effect of rpST administration was investigated in fattening pigs of three genotypes (Pietrain, crossbred between Dutch Yorkshire and Dutch Landrace (F1) and Duroc), slaughtered at 100 or 140 kg live weight. At both slaughter weights there was a significant improvement of daily gain, feed conversion ratio, backfat thickness and meat percentage. At 140 kg daily gain was much more improved in Pietrain and F1 than in Duroc. Genotype * treatment interaction was significant for backfat thickness; the fatter the genotype, the more response. No effect on meat quality was found.

Protein accretion as measured by N retention increased significantly for all three genotypes. Fat deposition, measured by energy retention, decreased significantly in rpST treated animals.

PERFORMANCE, HEALTH, GRADING AND MEAT QUALITY AT 100 AND 140 KG

Experimental

In a growth experiment 32 animals (four litters of four barrows and four gilts) of each of three genotypes (Pietrain, crossbred between Dutch Yorkshire and Dutch Landrace (F1) and Duroc) were used. A control and a treated group were formed in which genotypes, litters and sexes were equally represented. Animals were allotted at random regarding weight and received rpST from a live weight of 60 kg at a level of 4 mg per day. Half of the animals was slaughtered at 100 kg live weight, the others at 140 kg. Pigs were fed ad libitum a diet with 18% crude protein and 2162 kcal net energy per kg. Calculated digestible lysine content was 0.92%; digestible methionine + cystine was 0.55%. During the experiment abnormalities in health and condition were registered daily. Backfat thickness was measured ultrasonically the day before slaughter.

After slaughter commercial grading and detailed carcass evaluation according to a standard dissection method were carried out. Meat quality was studied for taste, odour and tenderness organoleptically and according to Warner-Bratzler (shear force). Furthermore the chemical composition of the meat was determined as well as percentages drip and cooking loss and colour.

Results

State of health of the pigs was satisfactory. No differences were observed between controls and treated groups.

Performance of the animals during the experiment, as presented in Kanis et al (1988a) and summarized in Table 1, reflected the normal gain and feed conversion in Dutch pig production.

Daily gain during the period from 60-100 kg was improved significantly for Pietrain, F1 and Duroc with 3.5%, 6.0% and 3.9% respectively. For the period from 100-140 kg these improvements were 31.7%, 27.4% and 2.7%. The much larger response to rpST treatment of the European pigs is remarkable (Fig. 1).

Feed intake was reduced for all three genotypes with 5.2%, 4.2% and 3.6% in the range from 60-100 kg. From 100-140 kg intake was increased with 8.0% and 8.3% in the Pietrain and the F1 groups, while intake was decreased in the Duroc group.

Feed conversion ratio was significantly improved with 8.2%, 9.9% and 8.4% in the range from 60-100 kg. From 100-140 kg response to treatment in the European pigs (19.0% and 15.8%) was again considerably stronger than in the Duroc group (6.5%) (Fig. 2).

Backfat thickness at 100 kg was significantly decreased with 6.5%, 13.1% and 20.1%, and at 140 kg with 18.3%, 20.0% and 27.9%. Rate of response was correlated with the level of backfat thickness; larger in the Duroc than in the European genotypes (Fig. 3).

Lean parts were significantly increased at 100 kg with 2.3%, 3.7% and 7.4% and at 140 kg with 8.8%, 7.9% and 9.4%.

Table 1. Least squares means of daily gain (DG), food intake (FI), food conversions ratio (FC), backfat thickness (BF) and lean parts (LP) in control animals (CON) by genotype and weight range, deviation (%) in rpST treated animals from control animals and significance of treatment (T), genotype (G) and treatment * genotype interaction (T*G).

Trait	Genotype Pietrain CON	rpST	F1 CON	rpST	Duroc CON	rpST	Pooled over Genotypes CON	rpST	Significance T	G	T*G
60 to 100 kg or at 100 kg (n=95)											
DG (kg/d)	.952	+3.5	.962	+6.0	.868	+3.9	.927	+4.5	*	***	NS
FI (kg/d)	3.01	-5.2	2.89	-4.2	2.93	-3.6	2.94	-4.4	**	NS	NS
FC	3.18	-8.2	3.02	-9.9	3.37	-7.2	3.19	-8.4	***	***	NS
BF (mm)	12.3	-6.5	13.7	-13.1	17.5	-20.1	14.5	-14.0	***	***	**
LP (%)	56.1	+2.3	57.8	+3.7	52.7	+7.4	55.5	+4.4	***	***	NS
100 to 140 kg or at 140 kg (n=47)											
DG (kg/d)	.670	+31.7	.693	+27.4	.765	+2.7	.710	+19.9	***	NS	NS
FI (kg/d)	3.11	+8.0	2.99	+8.3	3.39	-5.4	3.16	+3.3	NS	NS	NS
FC	4.71	-19.0	4.39	-15.8	4.45	-6.5	4.52	-13.9	***	NS	NS
BF (mm)	17.4	-18.3	18.8	-20.0	24.4	-27.9	20.2	-22.7	***	***	*
LP (%)	52.2	+8.8	54.2	+7.9	50.4	+9.4	52.3	+8.7	***	***	NS
60 to 140 kg (n=47)											
DG (kg/d)	.769	+22.7	.797	+20.6	.822	+1.0	.796	+14.5	***	NS	*
FI (kg/d)	3.05	+2.8	2.95	+3.1	3.20	-5.5	3.07	0.0	NS	NS	NS
FC	4.00	-16.6	3.74	-15.1	3.89	-6.0	3.88	-12.5	***	**	NS

NS Non significant   **   P < .01
*   P < .05   *** P < .001

An outline of least squares means for carcass grading data and organ weights are summarized in Table 2 (Kanis et al, 1988b). Data are corrected for differences in carcass weight. The repartitioning effect of rpST is reflected in a meatier carcass. Most of the corresponding parameters were, particularly at 140 kg, significantly better for the rpST treated groups than for the controls for all three breeds.

Weights of heart, kidneys and liver were significantly increased at 140 kg. At 100 kg this was only the case for kidney weight.

Meat quality characteristics were determined in 18 animals per treatment. As was presented by Kanis et al (1988c), the organoleptic characteristics tenderness, odour and taste did not show significant differences.
Protein, fat and moisture content as well as the physical characteristics, drip loss, cooking loss, shear force and colour, showed no significant differences either.

Fig. 1. rpST effect on weight gain from 60 - 100 and 100 - 140 kg live weight.

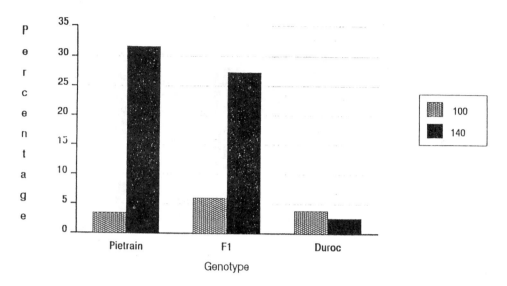

Fig. 2. rpST effect on kg feed/kg gain from 60-100 and 100-140 kg live weight.

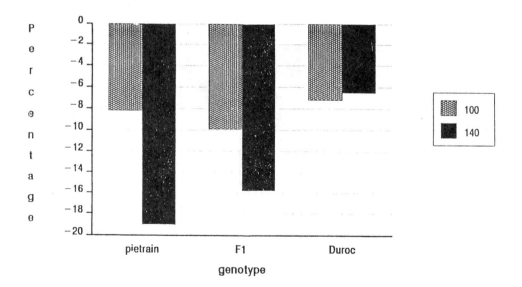

Fig. 3. rpST effect on back fat thickness at 100 kg and at 140 kg live weight.

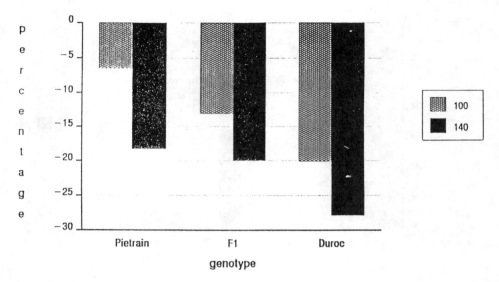

Fig. 4. rpST effect on repartitioning of nutrients from 85 - 100 kg live weight.

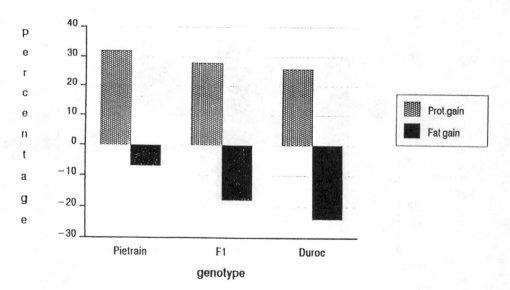

Table 2. Carcass characteristics and weights of some groups of muscles and organs by slaughter weight and treatment and significance of the treatment effect.

	Slaughter weight					
	100 kg			140 kg		
	Control	rpST	sig	Control	rpST	sig
Meat (%) HGP[1]	51.2	53.7	**	47.1	51.1	**
Conformation[2]	2.13	2.05	NS	2.37	2.06	*
pH ham	5.8	5.8	NS	5.7	5.8	**
Colour[3]	2.2	2.5	NS	2.7	2.5	NS
Ham (kg)	8.74	9.12	**	11.79	12.86	***
Loin (kg)	6.93	7.10	NS	9.61	10.29	***
Shoulder (kg)	4.45	4.70	***	6.10	6.63	***
Heart (g)	354	361	NS	439	478	***
Kidneys (g)	343	395	***	424	490	***
Liver (g)	1673	1745	NS	1919	2064	*

1) Meat percentage estimated according to the Hennesy Grading Probe
2) Conformation of carcass; scale: 1 (lean) to 4 (fat)
3) Japanese colour standards; scale: 1 (light) to 6 (dark)

PROTEIN AND ENERGY BALANCE EXPERIMENTS

Experimental

Nitrogen and energy balance experiments were carried out with twelve barrows from each of the mentioned genotypes.

Six animals of each genotype received rpST at a level of 4 mg per day from 60 kg live weight. The same diet was fed as in the performance experiment but in this case at a restricted level of 2.6 times maintenance requirement. Nitrogen retention was measured in individual metabolism cages for six successive periods of seven days each. The first period started 28 days after the first rpST administration. Faeces and urine were collected separately. Feed, faeces and urine were analyzed weekly for N, dry matter, organic matter and energy. Calcium and Phosphorus retention were determined for the second week only.

Energy balance was measured by placing the metabolism cages of each genotype during the 2nd and 5th period (P2 and P5) in two respiration chambers. Heat production was determined for two times 48 hours per period from continuous measurements of gaseous exchange of CO_2 and O_2. Energy content of feed, faeces and urine was determined by bomb calorimetry.

Results

The results are presented in Verstegen et al, (1989). A summary is shown in Table 3 and Fig. 4 for the test periods P2 and P5, during which N as well as energy retention were measured. Treatment effect was significant for gain of weight, protein and fat. Protein accretion was improved by 42, 44 and 30 grams per day per animal in Pietrain, F1 and Duroc. Fat deposition was in the same period decreased by 18, 64 and 47 grams per day per animal.

On average for all three genotypes protein accretion was increased by 40 g/d and fat deposition was decreased by 42 g/d, reflecting the average repartitioning of the nutrients by rpST administration.

Calcium and Phosphorus retention of controls and rpST treated animals is presented in Table 4. The data demonstrate a slightly increased retention for Ca and P in the rpST treated groups. The difference is significant for P only.

Table 3. Intake of metabolizable energy (kcal ME/kg$^{.75}$), weight gain, protein gain and fat gain in control (C) animals and rpST (T).

Breed	Period	Treatment	Intake ME kcal/kg.75	Weight Gain g/d	Protein Gain g/d	Fat Gain g/d
Pietrain	P2	C	258	783	144	233
		T	259	909	193	210
	P5	C	258	750	144	287
		T	258	840	180	275
F1	P2	C	263	857	188	242
		T	264	871	225	192
	P5	C	260	786	146	310
		T	257	889	197	233
Duroc	P2	C	259	548	116	252
		T	260	683	166	195
	P5	C	256	581	116	278
		T	261	667	126	245

Table 4. Apparently digested calcium and phosphorus in rpST and control animals. Amount of phosphorus and calcium excreted in urine as % of ingested and amounts retained (g/d) in the fifth week after administering in crossbred animals (Period P2). (between parenthesis SD.)

Treatment	Weight Initial kg	Weight Final kg	Digested Ca %	Digested P %	Urinary Excretion P %	Urinary Excretion Ca %	Retained Ca g/d	Retained P g/d
rpST	92.0 (5.4)	98.1 (5.5)	35.7	44.0	5.2	0.9	10.3 (1.7)	7.5 (0.8)
Control	87.6 (7.3)	93.6 (7.4)	31.2	39.2	5.2	1.0	8.7 (1.1)	6.3 (0.8)

REFERENCES

Kanis, E., K.H. De Greef, W. Van der Hel, J. Huisman, R.D. Politiek, M.W.A. Verstegen, P. Van der Wal and E.J. Van Weerden. Interactions of recombinant porcine somatotropin (rpST) treatment with genotype and slaughter weight for production traits of growing pigs, 1988a. J. An. Sc. 66(Suppl. 1): 296.

Kanis, E., W. Van der Hel, J. Huisman, G.J. Nieuwhof, R.D. Politiek, M.W.A. Verstegen, P. Van der Wal and E.J. Van Weerden. Effect of recombinant porcine somatotropin (rpST) treatment on carcass characteristics and organ weights of growing pigs, 1988b. J. An. Sc. 66(Suppl. 1): 280.

Kanis, E., W. Van der Hel, W.J.A. Kouwenberg, J. Huisman, R.D. Politiek, M.W.A. Verstegen, P. Van der Wal and E.J. Van Weerden. Effects of recombinant porcine somatotropin (rpST) on meat quality of pigs, 1988b. J. An. Sc. 66(Suppl. 1): 280.

Verstegen, M.W.A., W. Van der Hel, A.M. Henken, J. Huisman, E. Kanis, P. Van der Wal and E.J. Van Weerden, 1989. Effect of exogenous somatotropin on partitioning of nitrogen and energy in three genotypes of pigs. Submitted to J. An. Sci.

SERUM SOMATOTROPIN VALUES IN BOARS AND BARROWS, IMPLANTED WITH ANABOLIC STEROIDS AND COMPARED TO UNTREATED CONTROLS

R.O. De Wilde*, H. Deschuytere* & M. Corijn**

Veterinary Faculty University of Ghent
*Department of Animal Nutrition
Heidestraat, 19, 9220 Merelbeke
**Department of Reproduction and Fertility
Casinoplein, 24, B 9000 Gent

ABSTRACT

Serum somatotropin, nearly absent in castrated male pigs, can significantly be raised by oestradiol + testosteron implants and can result in leaner carcasses. In boars basal somatotropin values are higher than in barrows but more variable and steroids implants gave no consistent elevation of serum somatotropin.

Somatotropin has been proven to be a potential growth promotor in pigs (CAMPBELL et al. 1988, CHUNG et al. 1985). Growth promotion is more accentuated in restricted paired feeding, because in ad lib. feeding conditions the voluntary feed intake is depressed (CAMPBELL et al. 1988). The extra growth is mainly achieved by increased protein and water retention and decreased fat accretion.

Somatotropin needs to be injected within short intervals and is therefore cumbersome in large fattening units. Almost all belgian pig farmers apply the ad lib. system in which the effect of porcine somatotropin (pST) is less pronounced. In addition, the injection frequency is also a restricting factor for general application.

We tried out some alternative ways of stimulating the endogenous production of pST by comparison of entire males vs castrates treated or untreated with implanted anabolic steroids and by evaluating the halothane gene (stress susceptibility) in pST production. Boar fattening is known to produce leaner carcasses; implantation with steroids has the same effect (DE WILDE & LAUWERS, 1984) and the halothane gene also favours a lean carcass (DE WILDE, 1984). These pST determinations were taken in the concept of a series of experiments in which the above mentioned parameters and their effect on leaner carcasses were investigated under restricted feeding conditions in order to eliminate the variable feed intake level.

MATERIALS AND METHODS

In experiment 1 only castrated males (halothane positive and halothane negative) groups were used. One half was implanted at the live weight of 50 kg with 20 mg oestradiol + 200 mg testosterone[1]. Before implantation and every week after implantation blood samples were taken to examine levels of oestradiol-17-beta, testosterone and pST.

In experiment 2 only entire males, divided into halothane positive and halothane negative groups were utilized. At 50 kg 3 groups were formed within each halothane group : one group received a silicone rubber implant with 24 mg oestradiol[2], a second one was treated with a pelleted implant of 20 mg oestradiol + 200 mg testosterone[3] and a third group was left untreated. As in the first experiment blood samples were taken before implantation and at regular intervals thereafter.

Porcine somatotropin was measured by the RIA-technique (MARPLE & ABERLE, 1972) in experiment 1 and by the ELISA-technique with the first antiserum obtained in goats in experiment 2. Comparison of the 2 techniques gave a correlation of $r = 0.99$.

RESULTS AND DISCUSSION

The control barrows in exp. 1 (fig. 1) had near zero values with little variation. On the other hand, control boars in exp. 2 (fig. 2) had higher levels with greater variation. The variation is important even within boars, as was confirmed in another experiment, in animals with permanent cannulaes in order to measure the diurnal variation. Control boars and barrows hold the same values during the period 50-95 kg (42-56 days).

The effect of implantation is higher in barrows than in boars in absolute as well as in relative figures. Some values reach a level as high as was obtained by daily injections of pST (CAMPBELL et al. 1988, CHUNG et al. 1985).

In boars the oestradiol alone seems more effective than the combination oestradiol + testosteron. The difference is respons between barrows and boars was reflected in a more pronounced increase of the carcass leanness in barrows than in entire males (DESCHUYTERE et al. 1987).

---

[1] Implix BF, Roussel Uclaf, France
[2] Compudose 200, Elanco
[3] FTO, distributed by Upjohn Co.

The response of the implantation lasted in all cases for at least 1 month. At slaughter after about 50 days the implants could be recovered in most animals.

There was a clear increase in serum oestradiol and testosterone after implantation of the corresponding steroid in barrows although there was not net correlation between the steroid increase and the increase of pST. In boars steroid level of the implanted groups was never significantly different from the controls. Also for the serum steroid values there was a large diurnal variation in control and treated boars.

It was not possible to detect a systematic high level of serum hormones in the subsequent blood samples of some pigs and a low one in others so that the variation could not be explained by the presence of "high level" pigs vs "low level" pigs.

Finally, there was no clear effect of the halothane gene on the serum pST level, although the halothane positive animals has significantly leaner carcasses than the halothane negative pigs.

CONCLUSION

Alternative ways of enhancing endogenous somatotropin release is possible either by omitting castration of male pigs or by implantation of the non synthetic steroids oestradiol and(or) testosterone in castrated barrows. The respons is not guaranteed in all pigs, but some animals react with pST serum levels as high as in pST injected pigs.

ACKNOWLEDGEMENTS

This work was financially supported by IWONL Brussels. The porcine growth hormone needed for serum analysis was a gift from the "National hormone and pituitary program" (University of Maryland, School of Medicine, U.S.A.).
Dr. D.M. Marple, Dep. of Animal and Dairy Sci., Auburn University U.S.A. is sincerely thanked for supplying the pGH antiserum.
The technical assistance of Mr. De Rycke and Mr. E. Maes is gratefully acknowledged.

REFERENCES

CAMPBELL, R.G., STEELE, N.C., CAPERNA, T.J., MCMURTRY, J.P., SOLOMON, M.B. & MITCHELL, A.D., 1988 J. Anim. Sci. $\underline{66}$: 1643-1655
CHUNG, C.S., ETHERTON, T.D., WIGGINS, J.P. 1985. J. Anim. Sci. $\underline{60}$: 118-130

DESCHUYTERE, H., DE WILDE, R. & CORIJN, M., 1987. Med. Fac. Landb. Univ. Gent 52 : 1665-1672
DE WILDE, R.O. 1984. Livestock Prod. Sci. 11 : 303-313
DE WILDE, R.O. & LAUWERS, 1984. J. Anim. Sci. 59 : 1501-1509
MARPLE, D.N. & ABERLE E.D. 1972. J. Ani. Sci. 34 : 261-265

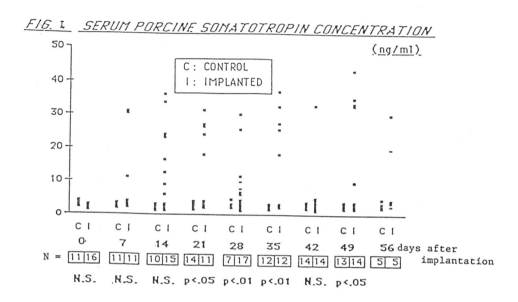

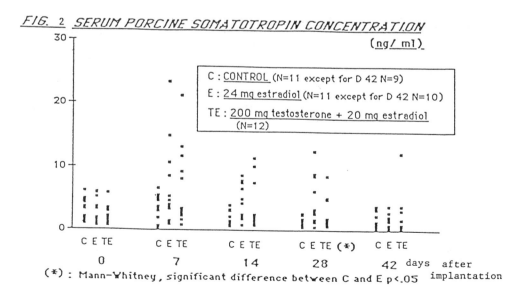

CONSEQUENCES OF DIETARY LIPID FEEDING IN PERIPARTURIENT SWINE ON ENDOGENOUS GROWTH HORMONE SECRETION AND SUBSEQUENT LITTER PERFORMANCE

G. Janssens and R. De Wilde
Laboratory of Animal Nutrition
State University of Ghent
Heidestraat, 19   9220 Merelbeke

ABSTRACT

Reproductive characteristics and endogenous growth hormone (GH) secretion were studied in sows receiving either a high or a normal lipid diet.
Except for a higher milkfat content, no clear improvement of evaluated performances in lipid fed sows was obvious. GH secretion proved to be higher in the fat supplemented group. Sows had lower GH concentrations in serum during pregnancy than during lactation. Most elevated GH secretion was noted in lactating sows after previous suckling of piglets.

INTRODUCTION

There is general agreement that high lipid feeding in gestating and lactating swine has beneficial effects on neonatal pig survival and performance (for review see Moser (1985), Pettigrew (1981)). An increased fat content in sows colostrum and milk seems to be the most consistent finding. Elevated glycogen stores in fetal liver and a higher carcass fat content are also reported. The exact mechanism though, of how hat feeding changes normal metabolic pathways is not fully established.
Steele et al. (1985) suggested that maternal periparturient endocrine adaptations in which the pituitary gland plays a keyrole, might be involved in the beneficial effect of lipid feeding. In his study, an increased prolactin secretion after thyrotropin-releasing hormone injection was noted in fat-fed sows. The dietary energy source (fat versus glucose) also influenced the serum insulin and GH concentrations.
The present study was conducted to determine the effect of a high lipid diet in late pregnant and lactating sows on reproductive characteristics and hormonal parameters such as GH, insulin and free fatty acids. Due to the purpose of this paper, results on GH concentrations in serum of sows will be extensively discussed, other results will be mentioned only briefly.

MATERIALS AND METHODS

Twenty-nine Belgian Landrace or crossbred sows of mixed parity were randomly assigned to one of two gestation-lactation diets containing either no added fat or 10 % fat as soybeanoil. The same diets were fed before and after parturition. Sows were placed on the experimental diets from day 100 of gestation and were restrictively fed until farrowing. From then the diets were provided ad libitum until the end of the third week of lactation. The two diets were isonitrogenous and met or exceeded all NRC (1979) requirements for nutrients. Due to the 10 % soybeanoil supplement, the diets differed highly in energy content as well as in energy source. Feed intake was recorded twice a week, water consumption daily. The sows were weighed every week. Number of pigs born, born alive, survival and weaned per litter was recorded. Piglets weights were recorded and milk samples were taken at birth, at 7, 14 and 21 days.

Immediately after birth, one piglet per litter was sacrificed for chemical analysis of total carcass fat content and liver glycogen. On day 109 of gestation 3 sows of the control and 6 of the fat supplemented diets were catheterized 24 hours before initiation of blood collection. The same was done on day 21 of lactation for 5, respectively 6 sows. The protocol for blood sampling was as follows :

day	physiological state	comment
110	gestation	sample collection by the following schedule : fasting = 0800, 0815, 0830 h feeding at 0830 h post-feeding = 0845, 0900, 0915 h post-prandial = 1130, 1145, 1200 h
21	lactation	piglets removed from the sow at 0800 h sample collection by the following schedule : fasting = 0830, 0845, 0900 h feeding at 0900 h post-feeding = 0915, 0930, 0945 h post-prandial = 1100, 1115, 1130 h suckling from 1330 h post-suckling = 1400, 1415, 1430 h

RESULTS AND DISCUSSION

Serum porcine GH was determined using the modified double antibody sandwich ELISA method measuring the alkaline phosphatase activity. A comparison between this method and earlier RIA technique revealed a correlation higher than 0.99. Endogenous GH secretion in sows fed high lipid versus control diets is presented in table 1.
Irrespective of dietary treatment and physiological state of sows, significant differences between the consecutive collection hours were found.
Basal (fasting) levels were lower than post-feeding levels. Post-prandial levels (3 h post-feeding, NB only 2 h in lac-

tation) returned to basal levels. Thirty minutes after suckling GH levels in sows were twice as high as basal levels. Dietary treatment resulted in elevated GH concentrations in soybeanoil fed sows. Basal levels in the latter were 18 % and 15 % higher during gestation respectively lactation. A more drastic increase was observed immediately post-feeding (31 % and 32 % respectively). Post suckling a comparable elevation could be detected (+ 28 %). Lactating sows had higher GH concentrations than pregnant sows. For both diets the augmentation was about 20 %. Compared to growing fattening pigs, the GH levels are in general much lower.

TABLE 1   Serum GH (ng/ml) in pregnant and lactating sows in relation to dietary energy source

		control n = 3	10 % soybeanoil n = 6	signif.
		$\bar{x} \pm$ SD	$\bar{x} \pm$ SD	
gestation d 110	fasting	1.09 $\pm$ 0.167	1.29 $\pm$ 0.107	$P<0.01$
	post-feeding	1.42 $\pm$ 0.194	1.86 $\pm$ 0.257	$P<0.01$
	post-prandial	1.16 $\pm$ 0.074	1.33 $\pm$ 0.184	$P<0.05$

fasting = mean value of blood samples collected at 0800, 0815, and 0830 h
feeding at 0830 h
post-feeding = mean value of blood samples collected at 0845, 0900 and 0915 h
post-prandial = mean value of blood samples collected at 1130, 1145 and 1200 h

		control n = 5	10 % soybeanoil n = 6	signif.
		$\bar{x} \pm$ SD	$\bar{x} \pm$ SD	
lactation d 22	fasting	1.29 $\pm$ 0.165	1.52 $\pm$ 0.269	$P<0.01$
	post-feeding	1.79 $\pm$ 0.340	2.37 $\pm$ 0.480	$P<0.01$
	post-prandial	1.29 $\pm$ 0.175	1.58 $\pm$ 0.342	$P<0.01$
	post-suckling	2.42 $\pm$ 0.267	3.10 $\pm$ 0.387	$P<0.01$

fasting = mean value of blood samples collected at 0830, 0845 and 0900 h
feeding at 0900 h
post-feeding = mean value of blood samples collected at 0915, 0930 and 0945 h
post-prandial = mean value of blood samples collected at 1100, 1115 and 1130 h
Pigs were allowed to suckle from 1330 h
After suckling mean value of blood samples collected at 1400, 1415 and 1430 h.

The present findings are conflicting with the results of Steele et al. (1985) in which no elevated serum GH levels occured post-feeding nor post-prandial in gestating swine. Also, no dietary treatment effect (glucose or cornoil supplement) was found in lactating sows. Nursing did not influence endocrine GH secretion in this experiment. Since soybeanoil was the only variable dietary factor between diets in our study, an endocrine adaptation of sows as a response to lipid feeding, resulting in elevated serum GH concentrations, must be assumed. The relative importance of the increased serum GH titre can be questioned due to the lack of performance improvement. It might be just a physiological adaptation. On the contrary, the higher milkfat content on day 14 of lactation, the tendency towards increased carcass fat and liver glycogen stores in neonatal pigs, a higher lipolytic rate in maternal backfat, all these phenomena might be partly the result of the translation of endocrine adaptations, triggered by dietary lipid, into specific metabolic effects in favour of lactating status.

ACKNOWLEDGMENTS

This work was financially supported by IWONL Brussels. The porcine GH for serum analysis was a gift from the "National hormone and pituitary program" (University of Maryland, School of Medicine, U.S.A.). The technical assistance of Mr. H. De Rycke and Mr. E. Maes are gratefully acknowledged.

REFERENCES

MOSER, B.D. 1985. The use of fats in sow diets. In "Recent developments in pig nutrition" (Ed. D.J.A. Cole and W. Haresign). p. 201-210.
PETTIGREW, J.E. 1981. Supplemental dietary fat for peripartal sows : a review. J. Anim. Sci. 53, p. 107-117.
STEELE, N.C., MC MURTRY, J.P. and ROSEBROUGH, R.W. 1985. Endocrine adaptations of periparturient swine to alteration of dietary energy source. J. Anim. Sci., 60, p. 1260-1271.
NRC., 1979. Nutrient requirements of domestic animals, no 2. Nutrient requirements of swine. Eight revised Ed. National Academy of Sciences-National Research Council, Washington DC.

FISH GROWTH HORMONES

A. Renard[*], C. Lecomte[*], F. Rentier[**], J.A. Martial[**]
[*]Eurogentec SA
Campus du Sart Tilman, B6, 4000 Liège, Belgium
[**]Genetic Engineering department
University of Liège, Sart Tilman B6, 4000 Liège, Belgium

At the time when the evolution of world fishery market was reaching a stationary phase, there has been, over the last ten years, a rapid increase in the extensive and intensive culture of various fish species. However, for salmonids, the growth of the industry is limited by various diseases which appeared with the intensive farming, by the cost of the food and by the length of the production cycle. It is thus desirable to introduce new treatments of disease, to increase the food conversion efficiency and the growth rate of fish. The latter has been attempted by selective breeding (a slow process), by increasing temperature (an expensive process), and by transfert to sea water. Anabolic steroids and thyroid hormones have been used but further studies are still needed to optimize treatment regimes and to meet the requirement of regulatory authorities. Recent interests have focused on the use of the growth hormone.

Fish possess a growth hormone (GH) which exhibits numerous biochemical and structural analogies with the mammalian GHs (Donaldson et al., 1979; Nicoll et al., 1986). It is produced by the somatotrops located in the pars distalis of the pituitary gland. Its secretion seems to be, as in mammals, under the control of two types of hypothalamic factors, one stimulating (GHRH like) and the other inhibiting (somatostatin like) (Jalabert et al., 1982).

Hypophysectomy results in cessation or a severe reduction of growth in various species which can be corrected by administration of exogenous GH. Moreover, numerous studies with mammalian GHs (reviewed in Donaldson et al., 1979) have shown that administration of these hormones to several fish species results in a significant increase in growth rate accompanied by increased appetite and improved feed conversion efficiency. The latter is probably mediated via a stimulation of fat mobilization and oxidation, a direct action on the rate of protein synthesis and/or breakdown and a stimulation of insulin synthesis and release (Markert et al., 1977). Higgs et al. (1978) showed that the bGH induced a greater increase in growth than the anabolic steroid methyltestosterone and the thyroid hormone thyroxine. It must be noted that although the mammalian GHs are active in fish. their efficiency is low since the growth promoting activity is only observed in the dose range of 1 to 100 μg of hormone per g of fish. It becomes thus clear that the mammalian GHs biological activities are distantly related to the ones of the fish hormone. Similarly it has been shown that the fish GHs are not very active in the rat tibia test and have a very low affinity for the mammalian hepatic receptors (Farmer et al., 1976). It is worth to mention that salmonid GHs are implicated in the control of adaptation to seawater and may influence the smoltification process (reviewed in Boeuf, 1988).

GH has been purified from several fish species i.e.: tilapia (Farmer et al., 1987), sturgeon (Farmer et al., 1981), carp (Cook et al., 1983), chum salmon (Wagner et al., 1985; Kawauchi et al., 1986), coho salmon (Nicoll et al., 1987) and eel (Kishida et al., 1987). Meanwhile, the GH cDNA of chum salmon (Sekine et al., 1985), yellow tail (Watabiki et al., 1986), rainbow trout (Agellon and Chen, 1986; Rentier et al., submitted), coho salmon (Nicoll et al., 1987; Gonzales-Villasenor et al., 1988), tilapia (Rentier et al., submitted) and tuna (Sato et al., 1988) have been cloned and expressed in E.coli. In chum salmon and in trout, the cDNAs of two forms of GH have been cloned (Sekine et al., 1985; Rentier et al., submitted). The trout and salmon growth hormones expressed in E.coli are able to efficiently promote growth in their respective species (Sekine et al., 1985; Agellon et al., 1988a). Moreover, these treatments were not significantly changing the chemical composition and ultrastructure of the fish (Agellon et al., 1988a). Recently, the GH gene of salmon (Agellon et al., 1988b) has been cloned and characterized thus opening the way to the development of transgenic animals.

Our laboratories are very active in this field and have cloned and expressed in E.coli and S.cerevisiae the growth hormones (and prolactins) of trout and tilapia. Our goal is to produce a hormone which could be economically used during the fish farming activities, not only to promote growth but also to influence osmoregulation. We are also developing new methods to deliver these proteins into the fish.

REFERENCES

Agellon, L.B. and Chen, T.T. 1986. DNA, 5, 463-471.
Agellon, L.B., Emery, C.J., Jones, J.M., Davies, S.L., Dingle, A.D. and Chen, T.T. 1988a. Can. J. Fish. Aquat. Sci., 45, 146-151.
Agellon, L.B., Davies, S.L., Chen, T.T. and Powers, D.A. 1988b. Proc. Natl. Acad. Sci. USA, 85, 5136-5141.
Boeuf, G. 1988. La pisciculture Francaise, 88, 5-21.
Cook, A.F., Wilson, S.W. and Peter, R.E. 1983. Gen. Comp. Endocrinol., 50, 335-347.
Donaldson, E.M., Fagerlund, U.H.M., Higgs, D.A. and McBride, J.R. 1979. Fish Physiology, 8, 455-597.
Farmer, S.W., Papkoff, H., Hayashida, T., Bewley, T.A., Bern, H.A. and Li, S.H. 1976. Gen. Comp. Endocrinol., 30, 91-100.
Farmer, S.W., Hayashida, T., Papkoff, H. and Polenov, A.L. 1981. Endocrinology, 108, 377-381.
Gonzales-Villasenor, L.I., Zhang, P., Chen, T.T. and Powers, D.A. 1988. Gene, 65, 239-246.
Higgs, D.A., Donaldson, E.M., McBride, J.R. and Dye, H.M. 1978. Can. J. Zool., 56, 1226-1231.
Jalabert, B., Fostier, A. and Breton, B. 1982. Oceanis, 8, 551-577.
Kawauchi, E., Moriyama, S., Yasuda, A., Yamaguchi, K., Shirahata, K., Kubota, J. and Hirano, T. 1986. Isolation and characterization of chum salmon growth hormone. Arch. Biochem. Biophys. 244, 542-552.
Kishida, M., Hirano T., Kubota, J., Hasegawa, S., Kawauchi, H., Yamaguchi, K. and Shirataki, K. 1987. Gen. Comp. Endocrinol., 65, 478-487.
Markert, J.R., Higgs, D.A., Dye, H.M. and MacQuarrie, D.W. 1977. Can. J. Zool., 55, 74-83.
Nicoll, C.S., Gregg, L.M., Russell, S.M. 1986. Endocrine Rev., 7, 169-203.
Nicoll, C.S., Steiny, S.S., King, D.S., Nishioka, R.S., Mayer, G.L., Eberrhardt, N.L., Baxter, J.D., Yamanaka, M.K., Miller, J.A., Seilhamer, J.J., Schilling, J.W. and Johnson, L.K. 1987. Gen. Comp. Endocrinol., 68, 387-399.
Sato, N., Watanabe, K., Murata, K., Sakagushi, M., Kariya, S., Nonaka, M. and Kimura, A. 1988. Biochem. Bioph. Acta, 949, 35-42.
Sekine, S., Mizukami, T., Nishi, T., Kuwana, Y., Saito, A., Sato, M., Ito, S. and Kawauchi, H. 1985. Proc. Natl. Acad. Sci. USA, 82, 4306-4310.
Wagner, G.F., Fargher, R.C., Brown, J.C. and McKeown, B.A. 1985. Gen. Comp. Endocrinol., 60, 27-34.
Watabiki, N., Tanaka, M., Masuda, N., Yamakawa, M., Nakajima, K. and Yoneda, Y. 1986. Sekagaku, 58, 986.

THE EFFECT OF BST ON PRODUCTION OF MILK INTENDED FOR THE PRODUCTION OF "GRANA CHEESE": TECHNICAL AND ECONOMIC ASPECTS

G. Piva, F. Masoero, A. Lazzari

Istituto di Scienze della Nutrizione
Universita' Cattolica del Sacro Cuore
Via Emilia Parmense 84
29100 Piacenza ITALY

ABSTRACT

In trials with BST on cows fed a diet for the production of Parmesan cheese, the authors demonstrated, with economical evaluations, that the net profit of the farm would be higher using BST in comparison with farms that do not produce milk for Parmesan cheese.

INTRODUCTION

A study we did on the introduction of BST on to the Italian market showed a certain drop in the number of cows (-200.000 between 1990 and 1995) (Piva, 1987), under a quota regime, even though it was assumed that the treatment would only be given to 30% of the animals in production.

Our work also showed that it would be the medium-size (about 50 heads) or medium large (70-120 heads) farms that would derive the greatest advantages from systematic use of the somatotropin.

The effect of the treatment would be economically advantageous mainly for those farms run according to the criteria of efficient management.

In Italy we find farms raising livestock that differ greatly from each other because they produce milk that will be used in particular ways:

1) farms producing milk that is to be sold directly as fresh milk, or that will be used in industrial cheese production (not typical or local cheese) and which are not restricted in their choice of feeding programme for the animals.

2) farms that produce milk that will be used in production of typical or local cheese (Parmesan, Grana Padano, Provolone, Emmenthal, Fontina etc.) which may feed cows with a restricted range of foodstuffs (silage is often excluded) and quantity is also limited.

More than 40% of the milk produced in Italy is used in this way.

Our research (Piva, 1988) carried out on cows fed a diet suited to production of "Grana Padano" cheese (hay, maize silage, concentrates) which lasted for a period of 3 months did not show appreciable differences as regards

TABLE 1: EFFECT OF BST ON MILK PRODUCED FROM COWS FED A DIET SUITABLE FOR PRODUCTION OF "GRANA PADANO" CHEESE.
(treatment every 4 weeks, 640 mg/head)

PARAMETER	DIFFERENCES BETWEEN TREATED AND CONTROL	
TREATMENTS MONTHS	10 (2)	3 (3)
FAT	none	none
CRUDE PROTEIN	none	none
LACTOSE	none	none
Ca, Na, K	none	none
P, Mg	----	increase
FREEZING POINT	none	----
pH	none	slight decreased
ACIDITY	none	increased
CREAMING	----	increased
SOMATIC CELL COUNTS	----	none
FIRMNESS OF THE CURD	----	none
BACTERIOLOGICAL AND RENNETING PROPERTIES	----	none
LACTODINANOMETRIC PARAMETERS	+ type E - F (A)	slight waisening
CHEESEMAKING PROPERTIES	----	none

(a) end of milking period.

TABLE 2: COMPARISON OF MILK PRODUCTION PROFITABILITY ON TYPICAL FARMS IN THE AREA OF "PARMESAN" CHEESE PRODUCTION AND ON FREE AREA.

AREA	FREE	"PARMESAN" CHEESE
PRICE OF MILK ECU/kg	0.40	0.53
WITHOUT B.S.T.		
NUMBER OF COWS	56.00	40.00
PRODUCTION PER HEAD kg/p.yr.	6500.00	5800.00
TOTAL PRODUCTION Tonn/p.yr.	364.00	232.00
EARNINGS PER HEAD ECU/p.yr.	2600.00	3074.00
COST OF MILK PER Kg (ECU)	0.36	0.40
COST PER COW (ECU/p.yr.)	2340.00	2320.00
PROFIT PER COW (ECU/p.yr.)	260.00	754.00
MILK PROFIT (ECU/kg)	0,04	0,15
WITHOUT TRANSFERABLE QUOTA		
WITH BST		
NUMBER OF COWS	49.00	35.00
PRODUCTION PER HEAD Kg/p.yr.	7429.00	6629.00
TOTAL PRODUCTION Tonn/p.yr.	364.00	232.00
EARNINGS PER HEAD ECU/p.yr.2	971.60	3513.00
COST OF MILK PER Kg (ECU)	0.35	0.40
COST PER COW (ECU/p.y.)	2600.00	2651.60
PROFIT PER COW (ECU/p.y.)	371.60	861.40
MILK PROFIT (ECU/kg)	0.05	0.14
DIFFERENCE IN NET PROFIT PER		
HEAD ECU/kg	111.60	107.40
DIFFERENT MILK PROFIT ECU/kg	0.01	- -

ASSUMPTIONS:
  Effect of BST on milk production : + 13.5%
  Cost of treatment (ECU/head/day) : 0.27
  Period of treatment (days) : 210.00

TABLE 3: COMPARISON OF MILK PRODUCTION PROFITABILITY ON TYPICAL FARMS IN THE AREA OF "PARMESAN" CHEESE PRODUCTION AND ON FREE AREA.

AREA		FREE	"PARMESAN" CHEESE
PRICE OF MILK	ECU/kg	0.40	0.53
WITHOUT BST			
NUMBER OF CATTLE		56.00	40.00
PRODUCTION PER HEAD	kg/p.yr.	6500.00	5800.00
TOTAL PRODUCTION	Tonn/p.yr.	364.00	232.00
EARNINGS PER HEAD	ECU/p.yr.	2600.00	3074.00
COST OF MILK PER Kg	(ECU)	0.36	0.40
COST PER COW	(ECU/p.yr.)	2340.00	2320.00
PROFIT PER COW	(ECU/p.yr.)	260.00	754.00
MILK PROFIT	(ECU/kg)	0.04	0.13
UNDER A REGIME OF A SINGLE NATIONAL POOL			
WITH B.S.T.			
NUMBER OF CATTLE		56.00	40.50
PRODUCTION PER HEAD	kg/p.yr.	7377.00	6583.00
TOTAL PRODUCTION	Tonn/p.yr.	413.00	263.20
EARNINGS PER HEAD	ECU/p.yr.	2950.00	3489.00
COST OF MILK	(ECU.kg)	0.32	0.37
COST PER COW	(ECU/p.yr.)	2360.00	2435.70
PROFIT PER COW	(ECU/p.yr.)	590.00	1053.30
MILK PROFIT	(ECU/kg)	0.08	0.16
DIFFERENCE IN NET PROFIT PER			
HEAD	ECU/p.yr.	330.00	299.30
DIFFERENCE MILK PROFIT	ECU/kg	0.04	0.03

ASSUMPTIONS:
  Effect of BST on milk production   : + 13.5%
  Cost of treatment   (ECU/head/day)   :   0.27
  Period of treatment   (days)   : 210.00

the physico-chemical, enzymatic, bacteriological and renneting properties of the milk (table 1).

On the basis of these data it seems that we could allow for the possibility of using milk produced by cows treated with BST for the production of "Parmesan" cheese, and so we have re-elaborated a hypothesis of the economic impact of BST on a typical dairy farm specialized in production of milk for such cheese (tables 2 and 3).

CONCLUSION

When the milk is to be used for the production of "Parmesan" cheese, the net profit of the farm would be higher using BST, especially with transferable quota but the advantage is less in comparison with the farms that do not produce milk for the "Parmesan" cheese.

REFERENCES

Piva, G., Agrobiotech '87, Bologna, 10-13/11/1987.
Piva, G, Battistotti, B., Bottazzi, V., Masoero, F., 1988, person. commun.
Bertoni, G., Battistotti, B., 1988, person. commun.

USE OF SOMATOTROPIN IN LIVESTOCK PRODUCTION IN THE EUROPEAN COMMUNITY:
SEMINAR SUMMARY AND CONCLUDING REMARKS

W.F. Raymond* and A. Neimann-Sørensen**
*Periwinkle Cottage, Christmas Common
Watlington, Oxon OX9 5HR, UK
** National Institute of Animal Science
Foulum, 8830 Tjele, Denmark

1 INTRODUCTION

It has been known since the 1930's that one of the main agents controlling growth and production in animals, including man, is a growth hormone, somatotropin, which is secreted by the pituitary gland. In general the activity of this hormone is species-specific so that, for example, bovine somatotropin (BST), secreted by the dairy cow, has no influence on human growth processes. It was also shown with a number of animal species that injection of a pituitary extract from the same species would stimulate growth and milk production, indicating that rate of production may be limited by the amount of hormone being secreted.

For many years this knowledge could not be applied in practice because pituitary extract could only be obtained post mortem and, even with the use of abattoir byproducts, the supply was only enough for limited experimental purposes. The situation was transformed in the 1970's with the development of recombinant DNA techniques, which allowed somatotropins, specific for different animal species, to be produced in large enough quantities for more extensive experiments with farm animals. These experiments have shown that, like their natural analogues, the recombinant somatotropins (r-ST) increase rates of animal growth and production. Most experimental work to date has been with dairy cows, and the use of bovine somatotropin (BST) in milk production is now close to practical application.

The possible introduction into farm practice of this new technique for increasing animal production raises a number of important questions, relating in particular to the likely form and scale on which the technique might be adopted; the probable economic and social consequences if it were adopted; and the acceptability of the technique to consumers and the wider society. There has been some debate on these questions in the USA, where most of the experimental work has been carried out. For a number of

reasons, however, these questions could be of greater significance within the European Community (EC) -
- Community output of several animal products is already either in surplus (milk and beef) or close to surplus (pig-meat). Thus the consequences of a further increase in output, resulting from the use of somatotropins, demand careful study. In the case of milk, whose total output in the EC is limited by quota, it is also most important to distinguish the effects of BST in raising milk yield per cow from its effects on efficiency of milk production (productivity).
- milk supports attracts a major part of Community spending on agriculture, and the likely effect of the use of BST on support costs must also be considered.
- sales of milk and beef together make up 60% of farm sales in the EC; they represent an even higher proportion of the incomes of small farms, which are a key component in the rural social structure of several Member States (although only 9% of EC dairy cows are in herds of less than 10 cows, these represent 45% of the dairy herds). Thus it is important to examine whether use of BST could significantly affect the competitive position of small vis-a-vis larger dairy farms.
- with the trauma of post-war food shortage largely replaced by the problems of food surplus, and with rising disposable incomes, European consumers are placing increasing importance on food quality and safety. In part this is expressed in the concern that quality may be impaired by 'intensive' conditions of production; in part that food resulting from some methods of production may be inherently unsafe. Thus public concern at the use of steroid hormones and other growth promoters in meat production (reinforced by some notorious examples of misuse) has led to the former being banned in the EC. While the chemical nature and the mode of action of the somatotropins differ completely from those of the banned steroid hormones, BST is a hormone, and the possibility that its use could impair the 'quality' and safety of milk has aroused concern among some consumers.
- increasing importance is also being placed on the need to safeguard the 'welfare' of the livestock kept on farms in the Community, with main concern directed at high-producing animals kept under intensive conditions of management. Many dairy cows are considered to be in this category; thus the use of BST to further increase milk output by the

dairy cow is being questioned on welfare grounds.

A considerable experimental programme, mainly with dairy cows, and much of it funded by the pharmaceutical companies who developed the recombinant somatotropins, began in several EC countries in 1985. One objective of this work has been to provide the regulatory authorities, at both national and EC level, with evidence to show that the milk produced is safe for consumers. This evidence has been accepted by several authorities, and bovine somatotropin (BST) is now undergoing extensive farm testing, prior to application from the companies for a 'commercial product licence' to allow it to enter general farm use. However, for the reasons noted above, the introduction of BST could have much wider impacts than the introduction of a new vaccine or antibiotic, for which the present regulations were originally framed. Yet neither at Community, nor at Member State level, do regulatory mechanisms seem to be available which can take into account such wider implications before a product licence is granted.

Aware of these wider concerns, the European Parliament in July 1988 requested the Commission and the Council of Agricultural Ministers to make a comprehensive study of the likely consequences that would result if BST were to be used in European agriculture (Resolution DOC.A2-30/88) - and, in para 18, also required that any decision on the issuing of a product licence for BST should take account of these wider issues, despite the absence of formal regulatory authority. In response the Commission (DG III and DG VI) convened a 3-day seminar in Brussels, from 27-29 September 1988, attended by workers from both state and commercial research centres and by senior members of the Community services, and charged with examining and reporting on these wider issues.

The following Sections record the discussions, and the conclusions reached, on a number of the key concerns raised by the Parliament -
- Consumer health and safety (para 3.3)
- Animal health (5)
- Product quality for human nutrition (3.2; 4)
- Product quality for milk processing (3.2)
- Animal welfare (5; 7.3)
- Economic consequences (6.1; 7.2)
- Socio-structural impacts, and effects on the rural environment (6.2; 7.2; 7.4)

The final Section (7) summarises the possible advantages and disadvantages that the introduction of somatotropins might have for EC agriculture, in a format which it is hoped will aid further discussion in the Commission, the Council, and the Parliament.

2 THE MODE OF ACTION OF SOMATOTROPIN

Somatotropin is a long-chain protein molecule produced in the pituitary gland of mammals. When it enters the blood stream it operates, either directly or via stimulation of another hormone, IGF1, to increase rates of bone and muscle growth and of production of milk components in the udder; the latter is achieved by a coordinated change in the metabolism of body tissues which alters the partitioning of nutrients in favour of milk synthesis. The increase in the supply of glucose for lactose production is met partly from increased gluconeogenesis and partly from reduced oxidation af glucose by other tissues. The extra milk fat is obtained either by reduced body-fat synthesis or by increased fat mobilisation, depending on the energy status of the animal. The extra protein is made available by an increase in nitrogen retention. Effectively then, when somatotropin is injected into an animal it reinforces the natural production and actions of somatotropin.

3 SOMATOTROPIN AND THE DAIRY COW
3.1 Milk production

The effect of BST on milk production were reviewed by Chilliard (FR). In the initial experiments with r-BST daily injections of 30 mg of the hormone were found to increase milk yield by up to 25%; there was no further response above this level. More recently prolonged-release formulations have been developed, so that injections only have to be made at 2 to 4-weekly intervals. Phipps (UK) reported a 20% increase in milk yield (3.7 kg milk per day) when BST was injected 2-weekly over a 32 week period. Thus, while the reduced frequency of injection reduces the risk of stress on the animal from the injections it does appear to give a somewhat lower milk response than daily injections; response also falls over the period after each injections, indicating an attenuation of the effect (Fig. 2, Phipps, UK). As already noted, BST operates by changing the metabolic processes in favour of milk production. Immediately following treatment the initial increase in milk output comes from body reserves, if the cow is in

negative energy balance. This is exactly what occurs in normal early
lactation, with the cow 'milking off her back'. This is then followed by a
marked increase in food intake as the cow begins to compensate for the
outflow of nutrients in the milk by eating more. In exactly the same way
the food intake of the BST-treated cow rises 4-6 weeks after the start of
treatment. During the remainder of the lactation intake can be as much as
10% above that of non-treated cows so that, by the start of the next
lactation, body-weight has been fully recovered (just as with the untreated
high-yielding cow). Because milk output rises more than food intake, and
because the animal's maintenance requirement is distributed over more
litres of milk (Lebzien et al., FRG) efficiency of food conversion to milk
can increase by up to 6%. Advantage must be taken of this improved feed
efficiency if use of BST is to show economic benefit.

3.2 Milk composition

A large number of tests have shown that BST has a negligible effect on
milk composition. While a slight rise in milk fat content and a fall in
protein on the days immediately following BST administration have been
reported, these have quickly disappeared. Overall any effects are
negligible compared with the natural variations in milk composition over
the lactation, at different seasons of the year, and as affected by
different feeding regimes. A detailed review by Van der Berg (NETH) also
found no effect of BST treatment on the suitability of milk for processing
- although it was suggested that some makers of specialist cheeses might
insist on non-treated milk (as makers of Emmenthal already refuse to accept
milk from silage-fed cows).

3.3 Milk quality and safety

Apart from these possible minor changes in milk composition, BST has
no effect on the nutritional quality of milk. However concern has been
expressed that BST will be transferred to the milk, and might present a
health hazard to humans. Exhaustive tests have now shown that BST levels in
'treated' milk are no higher than in 'normal' milk (with both being close
to the limits of present detection methods); BST is heat-labile, and most
of the hormone present in milk is destroyed during pasteurisation; any BST
which might survive this treatment would rapidly be digested to amino acids
in the human stomach; and even if a trace of BST were finally absorbed the

hormone has been shown to be inactive in humans. Administration of BST can lead to a small increase in the content in milk of the hormone IGF1, which, unlike BST, is also found in humans. However the levels found in milk have been well within the normal physiological range; further, IGF1, like BST, is a protein and is broken down during pasteurisation and digestion. Thus the present consensus is that BST-treated milk is nutritionally safe for human consumption.

In 2 out of 14 trials a marginal increase was found in the somatic cell count (scc) in the milk from cows treated with BST. Scc is now one of the main tests of milk bacteriological quality in the Community - in fact the standard is being tightened up, with cell count numbers above which milk will be rejected being reduced to 400,000/ml in 1989. Thus if the scc in milk from a herd which is using BST does increase, the economic consequences would force a tightening in management and hygiene. In a positive direction, there are indications that BST may benefit restoration of udder health following mastitis (Burvenich et al., B).

3.4 Reproductive performance

There is much evidence that, when BST is administered to dairy cows early in the lactation, breeding efficiency, as measured by conception rate and calving interval, is reduced. Exactly the same phenomenon is found in 'normal' dairy herds, with high-yielding cows in negative energy balance being more difficult to breed than lower-yielding cows. In theory this should be avoided by improved feeding; in practice it is most difficult to get the high-yielding cow to eat enough food in early lactation to avoid loss of body-weight. However with BST this problem can be mitigated by delaying treatment until after the cow has conceived; thus there is a very strong case for not starting injections until 80-100 days after calving. Although the increase in milk production will be less than if treatment were started earlier, the difference is likely to be small; intake also begins to increase after a few weeks so that the cow is only in negative energy balance for a short period. It should be noted that no increase in metabolic disorders has been reported as a result of BST treatment (see Section 5).

3.5 Dairy breeding programmes

Applied genetics has contributed greatly to the increased efficiency

of the EC dairy industry, particularly since the general adoption of
Artificial Insemination (AI), which has permitted the extensive use of
bulls of high genetic merit. Much of the success of these breeding
programmes has been based on on-farm recording of milk production. Thus
there is concern that the uncontrolled adoption of BST in commercial
practice could produce a spurious increase in the recorded genetic merit of
dairy bulls (Colleau, FR). Gravert (FRG) concluded that this would require
more daughters per bull to be tested to give equal accuracy in assessment
of the bull's breeding value. He also noted that, even if BST is recorded,
the test will be sensitive to the number of days after injection that
milk-recording is done, because the cyclic response to BST (Fig. 2, Phipps,
UK) could complicate the assessment of the breeding value of females.
However, because BST increases the yielding ability of the individual cow,
its use could allow greater weight to be given to important utility
characters such as ease of milking, better feet, and disease resistance,
than is the case with present breeding programmes.

4 SOMATOTROPIN AND MEAT PRODUCTION

Less research has been done on the use of somatotropin in meat
production than in milk production. Reports at the seminar by Enright (IRL)
on beef cattle and by Hanrahan (IRL) and by Henning et al. (FRG) on pigs
confirmed that recombinant bovine and porcine somatotropins can give
significant improvements in daily gains and feed conversion efficiency with
these two species. Probably the most significant result is that the
increased gains are mainly of protein (lean meat), with the result that
carcases from treated animals can contain 15% less fat than those from
untreated animals of considerable importance in relation both to consumer
preference for lean meat, and to health concerns over the consumption of
animal fat. As a result pigs and beef animals treated with somatotropins
(ST) can be taken to higher weights before they are marketed (important
with beef as the supply of calves will continue to decrease as the number
of dairy cows falls); with pigs less effluent is produced per kg of meat
produced, so reducing pollution risks. It appears, though, that the
technique is not yet ready for commercial application, partly because the
response has been more variable than with dairy cows (Pell et al., UK) and
also because to date it has required impractical daily injections with r-ST

(though a successful 6-week implant with pigs was reported, to replace the current daily injections). More also needs to be known about the nutrition of treated animals, in particular of protein supply to match the increased lean production.

5 THE HEALTH AND WELFARE OF ANIMALS TREATED WITH SOMATOTROPIN

There is now considerable evidence that the milk from BST-treated cows is a safe food for humans; thus increasing attention has been directed to the health and welfare of the animals that are treated. One cause for concern, the risk of animal distress from daily injections, has been at least partly obviated by the development of prolonged-release formulations, and every attempt must be made to further reduce the frequency of treatment, even at the loss of some efficacy. However, as the UK Farm Animal Welfare Council stated in December 1987 'while there is at present no evidence of any welfare problems... there are areas of uncertainty which could have welfare implications where additional scientific information may be required in order to take a more considered opinion.' Among these 'uncertainties' are -
- the effect of the time of commencement of BST treatment in relation to long-term reproductive efficiency and body protein and energy balance.
- possible skeletal and metabolic problems in heifers from BST-treated cows.
- possible cumulative effects of BST use on the same animal over a number of lactations.

The review by Phipps (UK) at the seminar reflected the priority that has been given to answering these key questions. A wide range of metabolic measurements made on treated animals has shown no evidence of disfunction (e.g. ketosis), provided feed intake is increased in line with increased milk production (Oldenbroek et al., NETH). As already noted, the slightly-reduced reproductive efficiency is similar to the problem often found with high-yielding cows in early lactation. When BST treatment has been delayed to 80 days post-partum there has been no effect on reproductive performance; thus there is a strong case for delaying treatment so as to avoid exacerbating this problem. No increased incidence of mastitis has been found even when BST has been used at higher-than-recommended doses. The small increase in somatic cell count recorded in some experiments confirms the need for strict udder hygiene. There has been some criticism that most of these studies have been short-term, and have not covered possible cumulative, long-term, effects. However data are now available from cows that are entering their fourth lactation treatment with BST, without any measurable clinical effects. Other studies have also shown that calves from BST-treated cows show no abnormal physical or physiological features; many heifer calves have been bred successfully and are now ready to enter milk production.

Concern about welfare per se has mainly related to possible injury or distress resulting from the injection of BST. Certainly the replacement of daily by 2- or 4-weekly injections has greatly reduced this risk. A key point for decision remains whether, if BST treatment is permitted, injections should be made by a veterinary specialist rather than by the herdsman. Clearly this, and other, welfare matters remain to be resolved.

6 IMPLICATIONS FOR FARM ECONOMICS AND MANAGEMENT
6.1 Milk production

Few results are yet available from the numerous large-scale farm investigations of BST treatment of dairy cows now underway. For a number of reasons, however, it seems that the 'benefits' in practice are likely to be less than from the many published experiments which, as Chilliard (FR) reported, have given increases in milk output of 20% or more -

- according to Bauman (USA), response to BST is sensitive to 'level of management' (Fig. 1, Phipps, UK); under most commercial farm conditions management is likely to be less rigorous than under the highly controlled conditions of experimental studies.
- many experiments have used daily injections of BST. In practice BST is more likely to be administered at intervals of 2 weeks or more, which is known to reduce the overall response.
- most experiments have studied the response to BST administered from about the 60th day of lactation. Because this can reduce breeding efficiency, and demands very careful feeding, first injection of BST is likely in practice to be delayed to 80-100 days - with consequent reduction in overall response.

Thus the response to BST in practice is likely to be nearer to the 9-12% range suggested by the Rennes group (Verite, FR) than the 20% found in experiments.

6.2 BST and costs of milk production

To achieve even this level of response will require careful attention to feeding to ensure that treated cows are able to eat the extra feed they need to match their increased milk output. It became clear from the discussion that in many research studies this extra feed has been provided in the form of concentrates. Recent economic studies suggest that this may not result in lower feed costs per litre of milk produced because the amount of concentrates fed per litre will not fall, and may even increase. Yet, under the quota conditions operating in the EC, feed costs must be reduced if dairy profitability is to be maintained. Thus an important observation is that BST increases the dairy cow's capacity to eat forage feeds, and it seems essential that as much as possible of the additional feed requirement should be provided by such feeds, which under most circumstances are cheaper than concentrate feeds. This could be of particular significance if it encourages milk production in 'grass-growing' areas of the Community.

The problem that milk quotas pose for a yield-enhancing technique such as BST was examined by Zeddies & Doluschitz (FRG). A farmer who uses BST to increase the average lactation yield of his cows by 10% has two options; to reduce his number of cows by 10%; or to buy in 10% extra quota. Quota-transfer regulations differ between the Member States; but in general buying-in quota to match the increased milk production from BST seems unlikely to be economic. Zeddies & Doluschitz concluded that the alternative strategy, of reducing herd numbers, is likely to be more practicable in large than in smaller herds. However whether this proves profitable will depend on a number of factors -

- the cost of the additional feed needed to cope with the increased milk output of the BST-treatet cows.
- the use made of the land released from milk production. It is generally assumed that less cows will use less land, and that the land released will be available for profitable alternative enterprises. But with the emphasis in the EC now on 'set-aside' rather than on more output the options for alternative land-use are becoming increasingly restricted. In many cases the preferred option must be to use the land released to grow as much as possible of the higher feed requirements of the BST-treated cows.
- one possible alternative enterprise would be beef production. However

if BST is used less calves would be born per litre of milk produced. This would make dairy-bred calves more expensive and so affect the economics of beef production from the dairy herd; equally it could make suckler-beef production more competitive (see 6.3).
- there are as yet few reliable estimates of the likely effects of BST on the 'fixed costs' of the business, in particular labour use.
- the manufacturers have so far given no indication of the likely cost of the BST treatment. This could have a big effect on the financial outcome, particularly in more marginal situations.

Thus it is not yet possible to give firm estimates of the likely scale of adoption of BST in EC farming. Zeddies & Doluschitz (FRG) reported one estimate that 20% of the cows in Germany would be treated. The general view of the seminar, however, was that the level of uptake would probably be less than this, and that on many farms BST was more likely to be used as a tactical management tool rather than as a routine treatment in milk production –
- to increase herd output towards the end of the quota year if production seems likely to fall below quota.
- to adjust the seasonality of production; in several countries seasonal price differentials are being widened so as to encourage a more level milk supply.
- to treat individual cows that respond well, rather than the whole herd.
- to shift milk production towards the later part of the lactation, when forage can make up a higher proportion of the ration – of particular advantage on smaller dairy farms which are often in grass-growing areas.

6.3 Beef production

Less is known about the possible effects of BST on the EC beef industry. Initially the first effects would probably be indirect –
- a short-term increase in cow-beef supply, as cows were culled from BST-treated herds.
- a subsequent fall in the number of calves available for beef production (offset by only a small fall in the number needed for dairy replacements). Almost inevitably this would increase the cost of beef

calves and so reduce the profitability of beef production. Only later, if BST were used on beef animals, might it be possible to recoup part of such higher costs by taking beef animals to higher slaughter weights.

7 BOVINE SOMATOTROPIN, COMMUNITY AGRICULTURAL POLICY, AND THE CONSUMER

Over the last 40 years dairy farmers in the EC have adopted a succession of new techniques, developed by both state and commercial research services, which have given a steady increase in the amount of milk produced by each dairy cow. These have included central breeding programmes and AI; improved grassland management; better hay and silage; more precise concentrate formulation; machine milking; better control of diseases and metabolic disorders. Initially yield increases were small, but the rate of increase suddenly speeded up in the early 1970's, as these separate developments began to be integrated into 'systems' of milk production (one reason that milk production is in surplus was that corrective action was not taken early enough because this surge in production was not detected soon enough).

Higher milk yields, resulting from new technology, have been an important factor in maintaining the profitability of dairy farming; more importantly, they have led to a steadily-falling 'real' price for milk and milk products to consumers. As an example, the average milk yield of the 3 million cows in the UK increased by 4% a year from 1972 until 1984 (when quotas were introduced); over the same period the 'real' consumer price for liquid milk fell by 20%.

7.1 The impact of milk quotas

In 1984 this impetus to higher yields was checked by the introduction of milk quotas. With a quota regime replacing the previous effectively unlimited market for milk the priority changed from producing more milk to producing milk as cheaply as possible. This meant reducing feed costs per litre of milk produced. Many farmers did this by feeding more forages and less concentrates, even though this reduced yields, as an alternative to keeping fewer but higher-yielding cows. In this way farmers in most EC countries - even though they are selling less milk than in 1984 - have managed to maintain their margins.

Much of the slack in production efficiency has now been taken up. Thus to remain profitable, particularly as quotas are further reduced, the EC dairy industry must continue to investigate, and to adopt, new techniques which will further reduce the cost of milk production - provided these techniques are seen to be of benefit to society as consumers, and acceptable to society as concerned citizens.

7.2 The effects of BST on the dairy industry

It is in this context that BST must be evaluated.

(1) BST is not a radical new discovery which will revolutionise milk production, but one of the continuing series of developments which have steadily improved the efficiency of milk production over the last 40 years. Put into context, even if every dairy farmer in the UK increased his milk output per cow by 12% by using BST, this would be equivalent only to the gain that was made by adopting other techniques during any 3-year period between 1972 and 1984.

(2) The advantages of a 'yield-increasing' agent are less evident in a quota situation than in an open market. BST should only be adopted by the individual farmer if it allows him to produce a litre of milk more cheaply. As was argued in 6.1, this could mean that uptake of BST would be limited. Overall its effect on EC milk production would be likely to be well below the 2 1/2% increase in milk output per cow recorded every year in the early 1980's.

(3) Thiede has predicted that, overall, adoption of BST could speed up the rate of decline in dairy cow numbers in the eight main milk-producing Member States by about 10%, with cow numbers falling by 5.4 million by 1995, compared with 4.9 million if BST is not used (Nagel, CEC). Because routine treatment with BST would be more practicable on larger than on smaller dairy farms (6.2), much of this additional fall in cow numbers would be on smaller farms. However this would represent only a small increase in the present rate of loss of small farms.

(4) On the other hand BST would be of particular economic value in helping small dairy farmers to adjust their milk output to quota, and to take advantage of seasonal price differentials (6.1).

(5) Such use of BST, if it led to a more level output of milk, should improve the economy of milk processing factories.

(6) In general BST would only be used in situations in which it would

reduce the cost of milk production. This could limit the scale of uptake, and so the effect on the price of milk. Overall, though, consumers should benefit - as they have from the many other technical advances that have been made in milk production.

7.3 The welfare issue

As noted in Section 5, there is now much evidence that BST does not affect the quality of milk for human consumption, and that the milk produced is fully suited for manufacturing purposes (3.2). If this is the case then the decision on whether the use of BST on EC dairy farms should be permitted is seen to depend mainly on ethical considerations -
- does BST cause damage, either short or long-term, to the health or well-being of the dairy cow or its offspring?
- does the use of BST impair the 'welfare' of the dairy cow?
- even if 'proper' use has been shown not to damage health or welfare, what is the safety margin in the use of BST?

These are the questions to which priority is now being given. For unless clear evidence can be provided that BST causes no harm to animal health or welfare, consumer resistance could result in a fall in the demand for milk and dairy products which could more than outweigh any economic advantage resulting from the use of BST.

The indications to date are that, provided it is properly used, BST will be given a clean 'bill of health'. If this is confirmed in the forthcoming official reports on product safety and animal health, the main argument against recombinant BST would appear to be that it is an 'unnatural' agent. Several delegates at a recent 'closed' conference of the International Dairy Federation, concerned that this would damage milk sales, proposed that there should be a world-wide ban on the use of BST. Such a ban seems most unlikely; yet without it a unilateral ban on the use of BST within the Community would make EC dairy products less able to compete on the world market. Thus, if the necessary health and welfare conditions are satisfied, the preferred course would appear to be for the use of BST to be authorised for milk production in the Community, but only under strictly controlled conditions.

7.4 A regulated use of BST

These conditions might include -

- only sustained-release preparations could be used, and daily injections would be prohibited (this would be possible to monitor, as sites of frequent injections can be detected).
- manufacturers and suppliers of BST would only be permitted to sell to 'registered' dairy farms; to qualify for the register a farm must have obtained advice from a veterinary surgeon and a dairy nutritionist on the correct administration of BST and on changes needed in the feeding and management regimes.
- each 'registered' farm would have prescribed veterinary visits to check on animal health and welfare. (Resolution DOC.A2-30/88, para 25, proposed that BST should only be administered by a veterinary specialist. Such a requirement would reduce the economic benefit from using BST, particularly in smaller dairy herds, in which treatment costs 'per head' tend to be higher than in larger herds, and must thus be questioned).
- dairy farmers would be actively advised to delay administration of BST until after the cow had conceived.

It is accepted that no system of supervision could completely prevent mis-use of BST. But it is considered that mis-use under a system such as that proposed would be small compared with that likely if a drug such as BST, which would be readily available in other countries, yet is virtually undetectable in use, were banned.

Finally, the debate over BST clearly has implications wider than just the decision as to whether the use of this particular agent should or should not be allowed. For the decision will be interpreted more widely as an indication of the Community's attitude to technical progress, in particular in biotechnology. Inevitably there will be occasions in the future when society, through its representatives in National Parliaments and in the European Parliament, will decide that a particular technical development is unacceptable, and that its application in the Community should not be permitted. BST can be seen as test case. Thus it is important, if a proper precedent is to be set for the similar debates of the future, that any decision that is taken on the use of BST should be seen to have combined rigorous scientific objectivity with a careful evaluation of the wider social and ethical concerns of society.

LIST OF PARTICIPANTS

BELGIUM
V. BIENFET
Federation of Veterinarians of the EEC
128, Avenue Albert
1060 Brussels

C. BURVENICH
Department of Veterinary Physiology
State University of Ghent
Casinoplein 24
9000 Ghent

D. DeBRABANDERE
Rijksstation voor Veevoeding
Scheldeweg 68
9231 Gontrode

R. DE WILDE
Laboratory of Animal Nutrition
State University of Ghent
Heidestraat 19
9220 Merelbeke

J. FABRY
Centre de Recherches Agronomiques de l'Etat
Station de Zootechnie
Chemin de Liroux
5800 Gembloux

G. JANSSENS
Laboratory of Animal Nutrition
State University of Ghent
Heidestraat 19
9220 Merelbeke

W.W. KRÖGEL
Conseiller Technique d'Agriculture
Groupe PPE, Parlement Européen
1040 Brussels

C. LECOMTE
Eurogentec SA
Campus du Sart Tilman, B6
4000 Liege

F. RENTIER
Genetic Engineering department
University of Liege,
Sart Tilman B6
4000 Liege

M. TRACEY
20, rue Emile Francois
1474 Ways-la-Hutte

W. VANDAELE
Federation of European Veterinarians
in Industry and Research
Avenue Fonsny 41
1060 Brussels

CANADA

D. PETITCLERC
Agriculture Canada
Research Station
Lennoxville
Quebec J1M IZ3

DENMARK

N. AGERGAARD
National Institute of Animal Science
Department of Animal Physiology
Foulum Research Centre
8830 Tjele

A. NEIMANN-SØRENSEN
National Institute of Animal Science
Department of Cattle and Sheep Research
Foulum Research Centre
8830 Tjele

K. SEJRSEN
National Institute of Animal Science
Department of Cattle and Sheep Research
Foulum Research Centre
8830 Tjele

M. VESTERGAARD
National Institute of Animal Science
Department of Cattle and Sheep Research
Foulum Research Centre
8830 Tjele

FRANCE

Y. CHILLIARD
Laboratoire de la Lactation
INRA
Theix
63122 Ceyrat

J.J. COLLEAU
Station de Genetique Quantitative et Appliquee
INRA
78350 Jouy en Josas

F. DELETANG
Responsable Scientifique-Santé Animale-SANOFI
Scientific Manager-Animal Health Division-SANOFI
B.P. 126
33501 Libourne

J. DJIANE
INRA
Departement Elevage et Nutrition des Herbivores
Theix
63122 Ceyrat

R. JARRIGE
INRA
Departement Elevage et Nutrition des Herbivores
Theix
63122 Ceyrat

M. JOURNET
INRA
Departement Elevage et Nutrition des Herbivores
Saint Gilles
35590 L'Hermitage

R. VERITE
INRA
Station de Recherches la Vache Laitiére
Saint-Gilles
35590 L'Hermitage

GREECE

G. ANASTASIOS
Agricultural Counsellor
Permanent Representative of Greece at EEC
Avenue de Cortenberg 71
Brussels

G. VEIMOS
Directorate of Veterinary Research
Ministry of Agriculture
Acharnon 2
Athens

IRELAND

P. ALLEN
The National Food Centre
Dunsinea, Castleknock
Dublin 15

W. ENRIGHT
Grange Research Centre
Grange, Dunsany
Co. Meath

T. HANRAHAN
The Agricultural Institute
Moorepark Research Centre
Fermoy
Co. Cork

J.F. O'GRADY
Grange Research Centre
Grange, Dunsany
Co. Meath

ITALY

R. ALEANDRI
Associazione Italiana Allevatori
Via Tommassetti 9
Roma

G. PIVA
Istituto di Scienze della Nutrizione
Universita Cattolica del Sacro Cuore
Via Emilia Parmense, 84
29100 Piacenza

A. ROMITA
Istituto Sperimentale per la Zootecnica - Roma
Via Salaria, 31
00016 Monterotondo Scalo (RM)

LUXEMBOURG

E. WAGNER
Administration des Services Techniques
de l'Agriculture
16 Route d'Esch
Luxembourg

NETHERLANDS

G. VAN DEN BERG
NIZO
P.O. Box 20
6710 BA Ede

G.J. GARSSEN
IVO "Schoonoord"
P.O. Box 501
3700 AM Zeist

J.K. OLDENBROEK
IVO "Schoonoord"
P.O. Box 501
3700 AM Zeist

W. SYBESMA
IVO "Schoonoord"
P.O. Box 501
3700 AM Zeist

P. VAN DER WAL
Wageningen Agricultural University
Bureau International Research
Ritzema Bosweg 32a
6703 AZ Wageningen

J.G. DE WILT
Directie Landbouwkundig Onderzoek
Mansholtlaan 4
Postbus 59
6700 AB Wageningen

PORTUGAL

J. MIRA
Estacao Zootechnica Nacional
2000 Vale de Santarem

A.F. NUNES
Estacao Zootechnica Nacional
2000 Vale de Santarem

SPAIN

J.A. MARTINEZ
Universidad de Navarra
Departamento de Fisiologia Animal y Nutricion
Pamplona

UNITED KINGDOM
N. CRAVEN
Monsanto
Chineham Court
Chineham
Basingstoke RG24 0UL

N.A. EVANS
Pitman-Moore Ltd.
Breakspear Road South
Harefield, Uxbridge
Middlesex UB9 6LS

D.L. FELLER
Lilly Research Centre Ltd.
ERL WOOD MANOR
Windlesham
Surrey
GU20 6PH

D.J. FLINT
Hannah Research Institute
Ayr
Scotland KA6 5HL

G.E. LAMMING
University of Nothingham
Faculty of Agricultural Science
Sutton Bonington
Loughborough LE12 5RD

J.M. PELL
Institute for Grassland and Animal Production
Hurley Research Station
Maidenhead
Berks SL6 5LR

R.H. PHIPPS
Institute for Grassland and Animal Production
Church Lane
Shenfield
Reading RG2 9AQ

C.G. PROSSER
AFRC Institute of Animal Physiology
and Genetics Research
Babraham
Cambridge CB2 4AT

W.F. RAYMOND
Perriwinkle Cottage
Watlington
Oxfordshire OX9 5HR

R.G. VERNON
Hannah Research Institute
Ayr
Scotland KA6 5HL

WEST GERMANY	F. ELLENDORFF
Institut für Tierzucht und Tierverhalten
Der Bundesanstalt für Landwirtschaft
Mariensee
3057 Neustadt 1

E. FARRIES
Institut für Tierzucht und Tierverhalten
Der Bundesanstalt für Landwirtschaft
Mariensee
3057 Neustadt 1

H.O. GRAVERT
Institut für Milcherzeugung
Bundesanstalt für Milchforschung
Postfach 6069
2300 Kiel 14

M. HENNING
Institut für Tierzucht und Tierverhalten
Der Bundesanstalt für Landwirtschaft
Mariensee
3057 Neustadt 1

B. HOFFMANN
Ambulatorische und Geburtshilfliche
Veterinärklinik
Justus-Liebig-Universität Giessen
Frankfurter Strasse 106
6300 Giessen

P. LEBZIEN
Institut für Tierernährung
Der Bundesforschungsanstalt für Landwirtschaft
Bundesallee 50
3300 Braunschweig-Völkenrode

D. SCHAMS
Institut für Physiologie
Süddeutsche Versuchs- und Forschungs-
anstalt für Milchwirtschaft
Technische Universität-München
8050 Freising-Weihenstephan

M. SCHÖPE
IFO - Institut für Wirtschaftsforschung
Poschingerstrasse 5
Postfach 860460
8000 München 86

D. SMIDT
Institut für Tierzucht und Tierverhalten
Der Bundesanstalt für Landwirtschaft
Mariensee
3057 Neustadt 1

J. ZEDDIES
Institut für Landwirtschaftliche Betriebslehre
Universität Hohenheim
Schloss Osthof Süd
7000 Stuttgart 70

<u>YUGOSLAVIA</u>	R. CMILJANIC Institute of Endocrinology, Immunology and Nutrition P.O. Box 46 11080 Zemun Beograd
	R. JOVANOVIC Agricultural Faculty 21000 Novi Sad Veljka Vlahovica 2
	T. TESIC Federal Comittee for Agriculture Beograd
<u>USA</u>	D.E. BAUMAN Department of Animal Science Cornell University Ithaca NY 14853
	R.L. STOTISH American Cyanamid Company P.O. Box 400 Princeton NJ 08540
<u>CEC</u>	J. CONNELL J. GAY H. GLAESER B. HOGBEN R. NAGEL CEC Direction General de l'Agriculture, DG VI 200 Rue de la Loi 1049 Brussels